Shapes in Evolution
Part Six
Sacred Animals in Paleolithic
Sculpture
Their Evolution in Proto-historic
and Historic Religions

SHAPES IN EVOLUTION SERIES
by Pietro Gaietto

PART I
PHYLOGENESIS OF BEAUTY

PART II
INTELLIGENT CELLS AND THEIR INVENTIONS

PART III
EROTISM AND RELIGION

PART IV
ANTHROPOMORPHIC PALEOLITHIC SCULPTURE

PART V
CATALOGO DELLA SCULTURA PALEOLITICA EUROPEA
COLLEZIONE GAIETTO

PART VI
SACRED ANIMALS IN PALEOLITHIC SCULPTURE. THEIR EVOLUTION IN
PROTO-HISTORIC AND HISTORIC RELIGIONS

PART VII
AN ICONOGRAPHY OF WESTERN RELIGIONS FROM THE
PALEOLITHIC TO PRESENT DAY

PART VIII
CONCETTUARIO DEGLI STILI
GIROVAGANDO PER L'ARTE

PART IX
HORSE AND WHEEL

PART X
DOG AND MAN

PART XI
ORIGIN OF MAN

PART XII
CACCIA E GASTRONOMIA

PART XIII
LA MOLLETTA PINZANTE

PART XIV
ASCE

PART XV
HAPPINESS
LIFE'S IMPORTANT MOMENT

Editorial coordination by Licia Filingeri
Third edition revised, corrected and translated in English October 2019
English translation by **Paris Alexander Walker**.
Copyright © 2019 Pietro Gaietto
ISBN 978-0-244-23838-4
All rights reserved. It is forbidden to reproduce the work or parts thereof by any means, including printing, photocopying, digitization on the Internet and e-books, except as provided for by copyright law.

©2019 Pietro Gaietto
gaietto@fastwebnet.it

PIETRO GAIETTO

SACRED ANIMALS IN PALEOLITHIC SCULPTURE
THEIR EVOLUTION IN PROTO-HISTORIC AND HISTORIC RELIGIONS

Translated from the Italian

by Paris Alexander Walker

Sculpture of a lioness with human body. Ivory. Aurignacian (32,000 years old). Found in Hohlenstein-Stadel cave in Germany. Ulm Museum. (Detail from Fig. 45, page 23.)
Wikimedia Public Domain ©2008 Gaura

Table of Contents

Introduction	5
1 Animal species of the glacial and interglacial eras	7
2 Typology of sculptures of sacred animals	8
3 Mammoth	10
4 Rhinoceros	12
5 Hippopotamus	14
6 Lion	18
7 Leopard or Panther	24
8 Horse	26
9 Elk	27
10 Goat	31
11 Bison	34
12 Bull	36
13 Bear	38
14 Dog	41
15 Seal	43
16 Bird	44
17 Fish	50
18 Snake	53

A representation of a lioness sculpted in terracotta, 1.3 inches high. Gravettian civilization. Found at Dolní Věstonice, Moravia, Czech Republic. Age: 18,000-10,000 years.

Introduction

During the Middle and Upper Paleolithic, humans were present in chronological succession in four different species that are now extinct – *Homo habilis*, *Homo erectus*, archaic *Homo sapiens* and *Homo sapiens neanderthalensis*. *Homo sapiens sapiens* was present in five varieties: Grimaldi type, Combe-Capelle type, Předmostí type, Oberkassel type, Cro-Magnon type, and perhaps others yet unknown. There were many crosses between different species of humans as a result of frequent migrations and this contributed to the formation of modern humans.

During the Middle Paleolithic all these human species manufactured stone sculpture.

The oldest stone sculpture was discovered at Olduvai Gorge in Tanzania and was dated to 1,700,000 years. It is believed to have been produced by *Homo habilis*, the first human. It represents a vaguely human head that is interpreted by many prominent paleoethnologists as a representation of *Homo habilis*.

The oldest sculptures of the Lower Paleolithic are mainly anthropomorphic. In the Middle Paleolithic the scultptures that represent animals increase in percentage and so also in the Upper Paleolithic. The processing technique was gradually refined partly due to the use of ivory.

As for the Paleolithic sculptures depicting mammals, most of the archaeologists who have described them have generically called them animals. For about 50 years I too used the term *"animal"*, however I preferred the term *"mammal"* when the species was not clearly distinguishable.

In this book I attempt to assign species names to mammals: mammoth, rhinoceros, hippopotamus, lion, leopard, horse, elk, goat, bison, bull, bear, dog, seal. I have not interpreted the species for birds, fishes or snakes; it's impossible and also not useful, even as a vague indicator.

My thesis is that these animals were depicted in sculpture because they were considered "sacred", although some species were used for food. In this book it is pointed out that for some of these species there has been a continuity of worship in religions into historic periods, which in some cases continues today. In this regard you are provided with a rich documentation of photographs and drawings of sculptures to demonstrate the evolution from the Paleolithic to the present day.

The documentation is completed by photographs of currently living animal species, considering that, for the purposes of depiction in sculpture, these modern species do not possess different features than the extinct varieties from the Paleolithic.

Pietro Gaietto

A bust of the Egyptian goddess Sekhmet with a lion head. Circa 1370 AD. Altes Museum, Berlin.
(Detail of Fig. 37, page 20.)
Copyright 2006 Captmondo, Gnu Free Documentation License, Creative Commons Attribution-Share Alike 3.0 Unported

I

Animal species of the glacial and interglacial periods

During the period of their greatest expansion, glaciers covered Scandinavia, a part of England, Germany and the Alps. As a result over half of Europe was uninhabited.

Throughout the Quaternary period, for about two million years, there were periods of glaciation in Europe and different climates, and interglacial periods with even warmer climates than our current one.

During interglacial periods over the last 500,000 years, European fauna was of the African type. The modern hippopotamus *(Hippopotamus amphibius)*, which still survives in equatorial Africa, was widespread from southern Europe to southern England. In Italy, thousands of specimens have been found, clear evidence of their abundant presence. The hippopotamus lived alongside the extinct *Elephas antiquus* and *Rhinoceros mercki*.

During glacial periods, fauna was of the cold climate type. Humans and animals were present in Europe in areas uncovered by glaciers, and it appears that at certain times the climate was also quite mild.

In colder periods during the Quaternary period, the great auk (*Alca impennis* or *Pinguinus impennis* L.) inhabited the coastlines of Spain and southern Italy. Their remains have been found in the Romanelli cave, in the municipality of Castro, in Brindisi province. Its form was similar to that of penguins of today (Fig.1). The great auk became extinct at the end of the 19th century.

The symbolic animal of the Ice Age was the woolly mammoth (Fig. 2). It had thick hair and its tusks were much larger than those of today's elephants. Humans hunted this pachyderm assiduously to feed themselves, to obtain furs with which to cover themselves and take shelter from the cold, and for ivory from which they produced sculptures and jewelry. Huts made of mammoth bones have been found in Eastern Europe. Mammoths died out in the fifth millennium BC in Siberia.

The woolly rhinoceros *(Rhinoceros tichorhinus)* (Fig. 3) was a companion of the mammoth, and specimens have been found together perfectly preserved in the frozen tundra of Siberia as well as in deposits of *ozokerite* (earth wax) in Starunia, in Galicia.

The typical animals inside the Arctic Circle were the polar bears and modern seals that live there today. The polar bear *(Ursus maritimus)* of the Quaternary is known from a skull found in the region around Hamburg, mentioned by K. A. Von Zittel.

I have no knowledge of finds of the Quaternary polar seal on European coasts, but if the great auk lived on the Mediterranean coast I presume there was also at the same time the seal (Fig. 4).

In 1880 the monk seal *(Monachus monachus)* was frequent on all Italian beaches, while today it survives only on some beaches in Sardinia.

The Quaternary moose *(Alces latifrons)* lived all across Europe and during cold periods it was common in Italy. The modern moose *(Alces machlis)* (Fig. 5) is still present in Scandinavia and Northern Siberia. It was however far more widespread in Europe a few centuries ago; its disappearance is not due to climatic change but rather to hunting.

Fig. 1 Fig. 2 Fig. 3

Fig. 1 Penguins at the South Pole resemble the extinct great auk (*Alca impennis* or *Pinguinus impennis* L.) which lived at the North Pole.
Fig. 2 Woolly mammoth *(Mammuthus primigenius)*. Royal British Columbia Museum, Victoria, British Columbia.
Copyright 2008 Tracy, Creative Commons Attribution - Share Alike 2.0 Generic license. Creative Commons Wikimedia.
Fig. 3 Woolly rhinoceros *(Rhinoceros tichorinus)*.
Copyright 2010 Huhu Uet, GNU Free Documentation License, Creative Commons Attribution - Share Alike 3.0 Unported

Fig. 4   Fig. 5

Fig. 4 A common seal of the North Pole, similar to the monk seal *(Monachus monachus)* of the Mediterranean. Seals *(Phoca vitulina)* along the Fano coast. Denmark.
Copyright 2005 Baldhur, Wikimedia Creative Commons Attribution-Share Alike 3.0 Unported
Fig. 5 Elk *(Alces machlis)* similar to the Quaternary Moose *(Alces latifrons)*. Museum of Natural History "Giacomo Doria"of Genoa, Italy.

2

Typology of sculptures of sacred animals

In Paleolithic sculpture, depictions of animals are of five types:
1) Bicephalic head of man and animal;
2) An artistic hybrid of man and animal;
3) Bicephalic head of two animals;
4) A head of an animal without a neck;
5) An animal head atop a horizontal body without limbs, or with partial limbs.
The bicephalic head does not exist in Nature; it is an invention. Likewise the artistic hybrid of man-animal is, for the proto-historic and historic periods, defined by archaeologists to be "fabulous", "monstrous", "demonic", and so on. Even the animal's head without a neck is not natural, therefore it can be considered a symbolic invention. The representation of an animal's head on a body without limbs or with partial limbs can be caused by the difficulty of sculpting the limbs or from the uselessness of doing this for purposes of worship. Even the Paleolithic Venuses (naked women) had no feet and often no hands or arms. Moreover, in lithic sculpture we don't find horns and tusks represented; even small mammoth-ivory sculptures are free of tusks and limbs.
In this book "Paleolithic sculpture" refers to that found across the whole of Europe, but I include the Egyptian Paleolithic, where lithic sculptures were made that represented heads of crocodiles.
In the sculptures of various proto-historic and historic civilizations, following the invention of new processing

techniques in sculpture, the use of new tools and the building of temples, the animal's head was represented along with a body, in both human and animal. The body can have a human head or be composed of different parts of animals, as seen in the following examples. The same type, with different shapes, is present in various ancient civilizations. In the Egyptian civilization there were many deities with animal heads and human bodies such as Sobek (Fig. 6), god of waters and Nile floods, which had a crocodile head. The Assyrian civilization (9th century BC) had among its numerous deities a type with a vertical human body and a fantastic animal head (Fig. 7). It comes from Kalah (now Nimrud), one of the capitals of the Assyrian Empire. With the Hittites (2nd millennium BC), among the many deities was a zoo-anthropomorphic combination consisting of a lion with wings and a human head on its neck (Fig. 8). The next sculpture comes from Karkemish (between present-day Turkey and Syria). Winged men and mammals are still present today in religious art.

We find in the Hindu religion many representations of animals with human bodies, all incarnations of Vishnu. Varaha is the third incarnation of Vishnu and has a boar's head and human body. This example is from Nepal, 16th-17th century AD (Fig. 9). In India in the Hindu religion cows are sacred animals and roam freely in the streets even in large cities: in the photo (Fig. 10), a cow with a little calf roams undisturbed through the streets of an Indian city. Moreover, being sacred, its flesh is not eaten.

The Lamb of God is also represented in the liturgy of the Christian religion in sculptural and pictorial form. It is not uncommon on Easter holidays to make symbolic lamb-shaped sweets from flour or almond paste and to prepare roast lamb with potatoes (Fig. 11).

Fig. 6

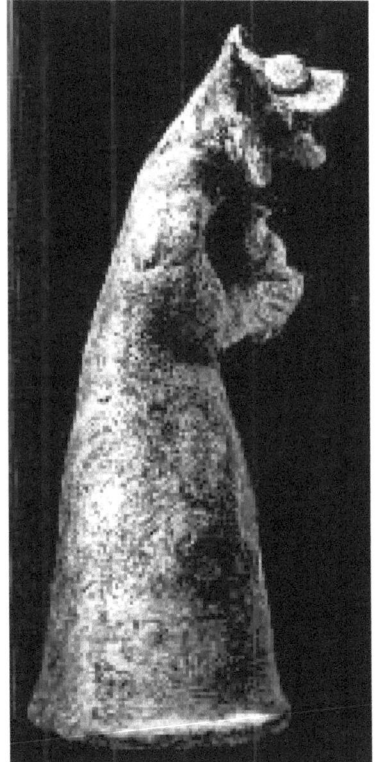

Fig. 7

Fig. 8

Fig. 6 Sobek, the Egyptian god with a crocodile's head.
Copyright 2009 Hedwig Storch, Wikimedia Creative Commons Attribution - Share Alike 3.0 Unported

Fig. 7 Small Assyrian terracotta sculpture, considered a "demon", representing a fantasy man with an animal head. Found at Kalak (Nimrud). Ninth century BC.

Fig. 8 Sculpture (bas-relief) of a Hittite divinity. It depicts a fantastic hybrid of Man-Lion and Bird.
Provenance: Karkemish (in present-day Turkey). Second millennium BC.

Fig. 9   Fig. 10   Fig. 11

Fig. 9 The boar-headed god Varaha of the Hindu religion, in a temple in Nepal.
Fig. 10 Hindu religion: a sacred cow in India.
Copyright 2007 Paris 75000 Public Domain Wikimedia.
Fig. 11 Christian religion. Divinity: the Lamb of God ("which takes away the sins of the world"). This small sculpture is made of almond paste. It is typical of Easter.

3

Mammoth

The woolly mammoth *(Elephas primigenius)* (Fig. 12) was smaller than the prehistoric elephant, reaching a height of about 11.5 feet. The rather sharp skull was tilted backwards. It had smaller ears than those of present-day elephants (Fig. 13). The tusks were huge, strongly curved and spiraled (see Fig. 2); each pair weighed an average of 220 pounds, but some were found that were over 13 feet long, each of which weighed 385 pounds.
It is a species of mammoth that roamed across Eurasia but, to my knowledge, in the protohistoric period there are no religious sculptures of men with mammoth heads, perhaps because the mammoth became extinct 5,000 years before our era. Moreover, it was a species that inhabited cold climates, while our first protohistoric civilizations arose in areas of warm climates.
In India, where elephants live and have lived, the Hindu god Ganesha (Fig. 14) has the head of an elephant and a human body.
It does not seem to me that Paleolithic sculptures depicting elephants (to the south) or mammoths (to the north) have been found in India, probably because they were not looked for, but they could be there, as the industries (along with the stone tools) found in those areas are as old as European and African ones, even if they exhibit slight differences.
The sculpture in Fig. 15 depicts a mammoth head with a sharp skull. On the right, at the end of the muzzle, where the trunk begins, there is the outline of a trunk bent to the side. In my previous publications I interpreted it as an *Elephas antiquus* head, because the technique for sculpting it is Acheulean, a cultural phase in which *Elephas antiquus* had not yet died out. It is flint, three and a half inches tall, and belongs to the Acheulean.
Fig. 16, according to an interpretation by paleoethnologists of the Czech Republic, represents a mammoth head and part of the body; the tusks are not represented. It is five inches long, carved in mammoth ivory. It belongs to the Upper Paleolithic (Gravettian).

tusks and limbs are not represented. It is carved in mammoth ivory and dates to the Upper Paleolithic (Gravettian).

The sculpture in Fig. 18, according to Russian paleoethnologists, also represents a mammoth head and body, and has no fangs or limbs. It is carved in stone, 0.7 inches long, from the Upper Paleolithic (Gravettian).

A small sculpture (Fig. 19) has been interpreted by French paleoethnologists as a mammoth. My interpretation is different, as it seems rather a bison with a massive head and body. It is carved on silicate stone. It was found outside any stratigraphic context. It is considered by French paleoethnologists as very similar in style to the mammoth sculptures found at Kostenki, Avdejevo and Předmostí.

Fig. 12

Fig. 13

Fig. 14

Fig. 12 An American mammoth (found in Utah) similar to the European mammoth of the Quaternary *(Elephas primigenius)*. Utah Museum of Natural History, USA.
Copyright 2009 Scott Catron, Creative Commons Attribution - Share Alike 2.0 Generic license.
Fig. 13 A present-day elephant.
Fig. 14 Ganesha, the elephant-headed god of the Hindu religion. Nepal, 17th century AD.

Fig.15

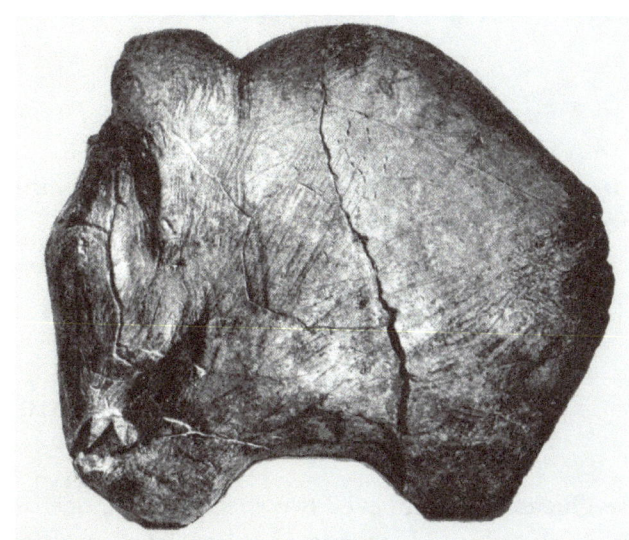
Fig. 16

Fig. 15 Sculpture of mammoth head found at Rodi Garganico (Foggia), Italy. P. Gaietto Collection.
Fig. 16 A mammoth head sculpture found at Předmostí, Moravia, Czech Republic.

Fig. 17

Fig. 18

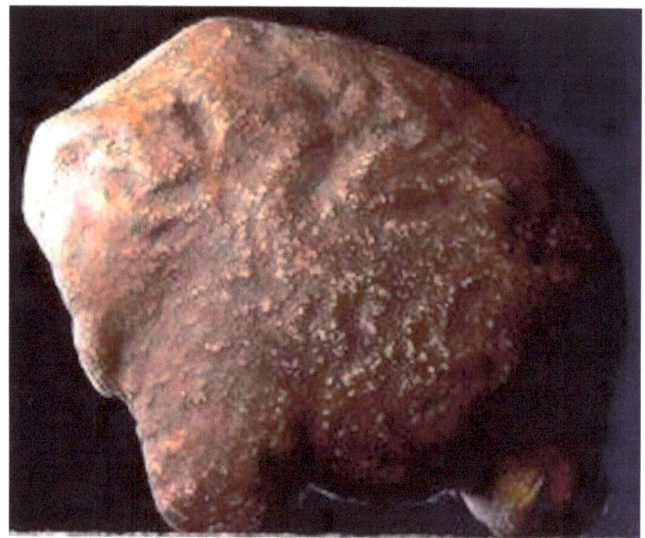
Fig. 19

Fig. 17 A mammoth sculpture found at Avdejevo, Republic of Karelia, Russia.
Fig. 18 A mammoth sculpture found at Kostenki, Russia. (From A. N. Rogacev.)
Fig. 19 A mammoth or bison sculpture found at Solutré, France. Departmental Prehistory Museum, Solutré, France.

4

Rhinoceros

In the Quaternary there lived two species of rhinoceros that are now extinct: the Merck's rhinoceros *(Rhinoceros mercki)*, coeval of *Elephant antiquus*, lived in Europe during warm periods, and the woolly rhinoceros *(Rhinoceros tichorhinus)* which appeared later and spread throughout Europe and Asia (see Fig. 3).
The woolly rhinoceros could attain a length of 12.5 feet and a height of 6.6 feet; it had two horns, the larger of which could be as long as 4 feet. The hair was at least 2.5 inches long, and long bristly hairs protruded from it. Currently rhinos are of various species, Asian (Fig. 20) and African.

In my previous books I considered the sculpture in Fig. 21 as a representation of a humanized animal that is an artistic hybrid between man and mammal, and I still consider it as such; if I wanted to name the animal, I think it could be a hybrid man-rhinoceros.

It is a head of a man-rhino with a neck, and without a horn, 5.5 inches high. It was found outside any stratigraphic context, but I consider its stylistic typology as Acheulean.

The sculpture in Fig. 22 is considered by Russian paleoethnologists as a representation of a rhinoceros. It is carved from stone and shaped by a sort of sanding. It is 1.1 inch long and dates to the Upper Paleolithic (Gravettian).

In Fig. 23 we see a rhino head modeled from burnt clay. Given the material and processing technique used, it was possible to depict the horn. Length 1.6 inch. Upper Paleolithic (Gravettian). Note the stylistic difference between this sculpture and the previous one (Fig. 22); both belong to the same cultural stage, called Gravettian.

Fig. 20

Fig. 21

Fig. 20 Black rhinoceros (*Rhinoceros unicornis*). India. Museum of Natural History "Giacomo Doria", Genoa, Italy.

Fig. 21 Artistic hybrid of man and rhinoceros. Varazze, Savona, Italy. P. Gaietto Collection.

Fig. 22

Fig. 23

Fig. 22 Rhinoceros sculpture found at Kostenki, Russia. (From A. N. Rogacev.)
Fig. 23 Sculpture of a rhinoceros head found at Dolní Věstonice, Czech Republic.

5

Hippopotamus

During the last interglacial period and its warmer climate, the hippopotamus (Fig. 24) was widespread throughout Europe and in particular in Italy. The cultural period was Lower Paleolithic, and the specific culture was Acheulean.
The Paleolithic sculptures I present here have all been found in Italy and are attributed to the Acheulean. They are of four types: hippopotamus head and human body; two-faced head of two hippopotamuses; single head of a hippopotamus; and two-faced head of a man-hippopotamus.
During the last glaciation, the hippopotamus disappeared from Europe but continues to be widespread in Africa where it became a sacred animal in Egyptian civilization. The goddess Thoeris was depicted in Egyptian sculpture with the head of a hippopotamus and the body of a pregnant woman (Fig. 25), and was particularly revered by pregnant women. In Fig. 26 we see an artistic hybrid of a hippopotamus head and a vertical human. This sculpture is 2.5 inches tall and carved from flint.
The sculpture in Fig. 27 (Photo: Silvano Maggi, 1961) is a fallen zoo-anthropomorphic menhir that depicts, like the previous sculpture, a hippopotamus head with a vertical human body. It is about 10 feet long. To photograph it, it was necessary to build a scaffolding in the trees. The light spots on the sculpture are rays of sunlight that filtered through the leaves.
In Fig. 28, a bicephalic zoo-anthropomorphic sculpture, 3.6 inches long. It is composed of a hybrid hippopotamus head (left) combined with a hippopotamus head. Acheulean. The sculpture of Fig. 29, somewhat damaged by alluvial rolling, depicts two heads joined at the neck, perhaps of a hippopotamus. Flint, 2.7 inches long. In Fig. 30 is a mammalian head, probably a hippopotamus. Length: 2.8 inches. The sculpture in Fig. 31 is bicephalic and zoo-anthropomorphic. It represents a human head (left) joined to another of perhaps a hippopotamus (right). It is in flint worked from all sides, with some wear from rolling. Length: 3.9 inches. The sculpture in Fig. 32 is bicephalic. It represents a hippopotamus head (left) joined to a human head. Flint, wornm, 4.1 inches in length. Acheulean. In Fig. 33, a human head (left) is joined to one of a hippopotamus. It is damaged by alluvial rolling, but recognizable. Length: 4.7 inches. Acheulean. The sculpture in Fig. 34 represents a head that looks like a hippopotamus (right) combined with one that looks human, with very worn features in the facial profile. The hippopotamus has a large eye (deliberately carved). Damaged by rolling. Length: 3.1 inches.

Fig. 24

Fig. 25

Fig. 24 Hippopotamus head (drawn for comics).
Fig. 25 An Egyptian deity: fertility goddess Thoeris with hippopotamus head and woman's body.

Fig. 26

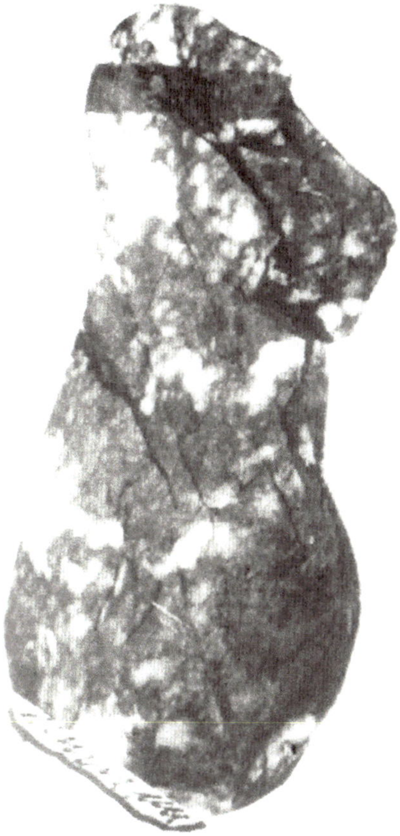

Fig. 27

Fig. 26 A sculpture with hippopotamus head and a vertical human body. Acheulean. Found at Rodi Garganico (Foggia), Italy. P. Gaietto Collection.
Fig. 27 A fallen zoo-anthropomorphic menhir having a hippopotamus head and vertical human body. Urbe, Locality Buschiazzi, Savona, Italy.

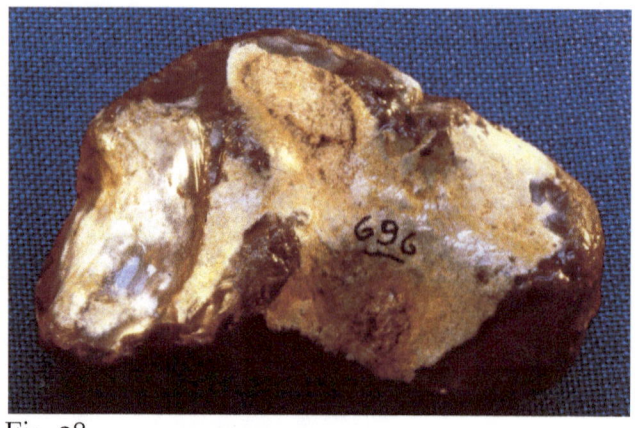

Fig. 28

SIDE VIEW

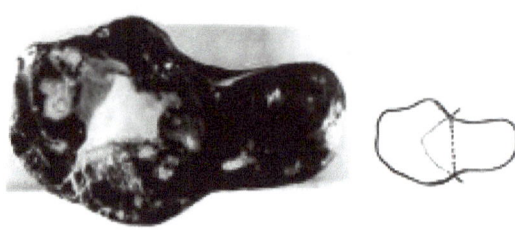

SEEN FROM BELOW

Fig. 28 Bicephalic zoo-anthropomorphic sculpture. It represents a hybrid head of a man-hippopotamus *(left)* combined with a hippopotamus head. Acheulean. Found at Rodi Garganico (Foggia), Italy. P. Gaietto Collection.

Fig. 29

Fig. 29 A sculpture slightly damaged by rolling. It represents two hippopotamus heads joined at the neck. Acheulean. Found at Rodi Garganico (Foggia), Italy. P. Gaietto Collection.

Fig. 30

Fig. 30 A zoomorphic sculpture showing a mammalian head, perhaps a hippopotamus. Flint. Found at Rodi Garganico (Foggia), Italy. P. Gaietto Collection.

Fig. 31

Fig. 31 A bicephalic zoo-anthropomorphic sculpture representing a human head *(left)* joined with a hippopotamus head. Acheulean. Found at Rodi Garganico (Foggia), Italy. P. Gaietto Collection.

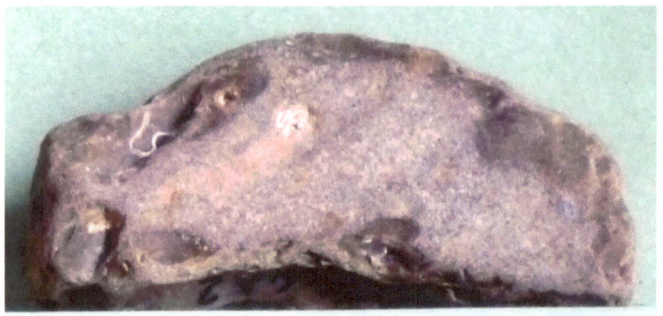
Fig. 32

Fig. 33

Fig. 32 A bicephalic zoo-anthropomorphic sculpture representing a hippopotamus head *(left)* joined to a human head. The rolling damage is not disfiguring. From Rodi Garganico (Foggia), Italy. P. Gaietto Collection.
Fig. 33 A bicephalic zoo-anthropomorphic sculpture representing a human head *(left)* joined to a hippopotamus head. Acheulean. Very damaged by rolling. Found at Torre in Pietra (Rome). P. Gaietto Collection.

Fig. 34

Fig. 34 A bicephalic zoo-anthropomorphic sculpture representing a hippopotamus head *(left)* joined to a human head damaged by alluvial rolling. Genoa, Sestri Ponente, Monte Gazzo. P. Gaietto Collection.

6

Lion

The cave lion *(Panthera leo spelaea)*, the largest of the Quaternary felines, lived throughout southern and central Europe up into the Magdalenian period. The common lion *(Panthera leo)* was present in the Balkan regions until the beginning of our era. The male lion (Fig. 35) has its head covered with a mane, which the lioness (Fig. 36) does not have.
In the Czech Republic, three lioness sculptures from the Upper Paleolithic (Gravettian) were found that were made with a new technique (two sculpted in burnt clay and one flat ivory carving). (See Figures 46, 47, 48.)
In every other Paleolithic sculpture in flint or other types of stone that used older techniques of processing from the Lower and Middle Paleolithic, it is not possible to distinguish the lion's sex since the mane was not depicted,

just as in mammoths the tusks were not represented. It is not to be excluded however that in the Paleolithic the sacred lion was female. In Egyptian civilization – one of the most ancient historic civilizations – the goddess Sekhmet, a solar deity, had a lion's (or lioness) head and a female human body (Fig. 37); she is shown naked with a youthful body and two beautiful breasts.

From the Gravettian – one of the last cultural phases of the Upper Paleolithic – one can hypothesize an evolutionary line with the Egyptian religion, since the two civilizations are only separated by a few millennia and it is known that a spiritual culture lasts longer than the material one. Every religion assigns the lion a name of a different divinity and different powers, and this is also true for the anthropomorphic divinities.

In Hindu religion, on the other hand, the god Narasimha, fourth incarnation of Vishnu, has a lion's head and a male human body (Fig. 38); he is endowed with different powers than those of the Egyptian goddess Sekhmet.

The bicephalic Paleolithic sculptures, that is two heads joined at the neck, all have a religious function and I hypothesize that even the single lion heads had the same purpose, and in my opinion this excludes the decorative function, that is to say, art for art's sake.

The sculpture in Fig. 39 is bicephalic zoo-anthropomorphic. It represents a lion's head (on the left) joined to a human one. Length: 5.1 inches. Final Acheulean or Mousterian of Acheulean tradition. Another bicephalic-faced zoo-anthropomorphic sculpture is shown in Fig. 40. The shape is clear, but the workmanship cannot be seen; this is due not only to type of stone used but also wear from rolling in streams. It represents a human head (on the left) joined to one resembling a lion's. The length is 4.7 inches.

A lion's head (left side) is shown in Fig. 41. The snout occupies two-thirds of the head. It is damaged by rolling, with partially erased traces of workmanship, but the shape is evident. Flint, from the Acheulean.

The sculpture in Fig. 42 depicting a lion's head was found in a large cave. The feline has a roaring expression with jaws wide open, adding drama to the style of work. It is a hanging sculpture; in fact it has four holes that allowed the passage of ropes to hang it. Other similar types have been found and were attributed to the Aurignacian; these are now exhibited in the Museum of Prehistory of the Troglodyte Fort near Sergeac, Dordogne (France). There are two vaguely geometric and zoomorphic sculptures considered as hanging sculptures by French paleoethnologists. The lion's head (Fig. 42) is 12.2 inches high; the four drawings show it hanging and with a view from all sides. I think it is Mousterian, given that Mousterian industries were found at the site. In the photo (Fig. 43) Licia Filingeri holds the sculpture to highlight its rear shape and elegance of style.

The sculpture in Fig. 44 depicts a lion's head without a neck. It is shown in semi-frontal and lateral views. I believe that this zoomorphic type, like the anthropomorphic sculptures, belongs to the Mousterian of the Acheulean tradition. It is 2.7 inches long with some slight rolling wear.

The Upper Paleolithic was the last cultural phase of the Paleolithic and, if compared to the Lower and Middle Paleolithic which lasted about two million years, it was relatively short, having a duration of only 25,000 years (from 35,000 to 10,000 years BC). In the Upper Paleolithic man began to use new materials to sculpt. He invented other processing techniques (as the older ones developed) and began producing sculptures exhibiting a new type of composition which foreshadowed the art of historic periods.

The sculpture in Fig. 45 is an example of the progress made in the Upper Paleolithic, about 32,000 years ago, during the Aurignacian. It depicts, according to some interpretations, a man with a lion's head; according to others it is a lioness with a human body. My interpretation is that the body is female, due to the shape of the legs. This sculpture had a religious function, like others of the historic epochs (Figures 37, 38 etc.). It also reveals the natural beauty together with the artistic sense, as beautiful legs have always been a source of attraction and highlighted as naturally beautiful. This sculpture is in mammoth ivory. It is the only Paleolithic sculpture having feet.

The sculpture in Fig. 46 represents a lioness on flat ivory suggesting remarkable body movement. It is from the last phase of the Upper Paleolithic (Gravettian).

Here's a lioness head (Fig. 47) modeled in burnt clay. Length is 1.5 inches. Gravettian.

The sculpture in Fig. 48 depicts, like the previous one, a lion's head and has a hint of a neck. It is modeled in burnt clay. The length is 2.3 inches. Gravettian.

Fig. 35

Fig. 36

Fig. 35 Asian lion (*Panthera leo persica*). Male.
Copyright 2011 Mousse, GNU Free Documentation License, Creative Commons Attribution - Share Alike 3.0 Unported.
Fig. 36 Lion (*Panthera leo*). Female.

Fig. 37

Fig. 38

Fig. 37 The Egyptian goddess Sekhmet with a lion's head. Circa 1370 BC. Altes Museum, Berlin.
CC BY-SA 3.0 © 2006 Captmondo
Fig. 38 Narasimha, a lion-headed Nepalese god of the Hindu religion. Twelfth to thirteenth century AD.

Fig. 39

Fig. 39 Bicephalic zoo-anthropomorphic sculpture representing a lion's head (on the left) joined with a human head. Périgueux (France). P. Gaietto Collection.

Fig. 40

Fig. 40 Bicephalic zoo-anthropomorphic sculpture representing a lion's head (on the left) joined with a head that looks like a lion. Found at Capo Rossello near Realmonte (Agrigento), Italy. P. Gaietto Collection.

 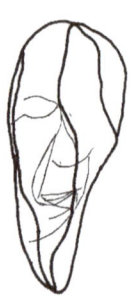

Fig. 41

Fig. 41 A zoomorphic sculpture representing a lion's head with the muzzle facing left. Very damaged by rolling. Found along the Alento river, Chieti, Italy. On the right is a frontal drawing of Fig. 41. P. Gaietto Collection.

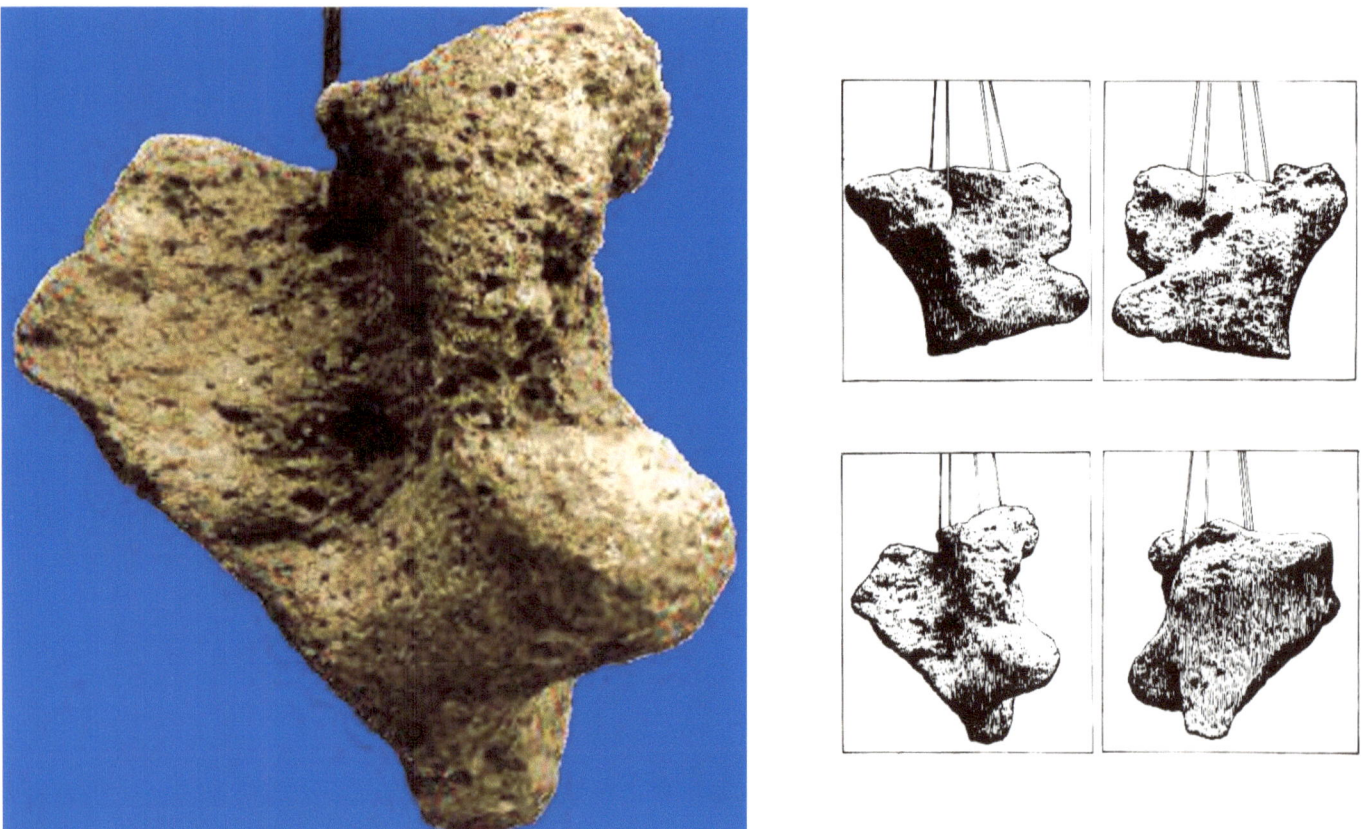

Fig. 42

Fig. 42 Zoomorphic sculpture representing a quite expressive head, interpreted as a roaring lion. It is sculpted for hanging. Mousterian. Found at Grotta delle Manie, Varigotti (Savona), Italy. P. Gaietto Collection.

Fig. 43

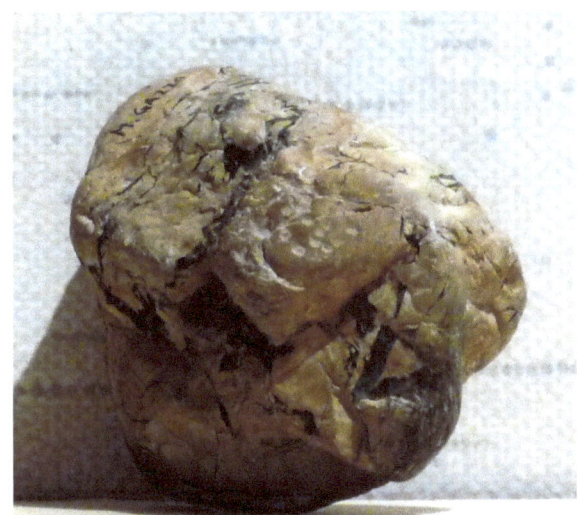
Fig. 44

Fig. 43 Licia Filingeri with the lion's head, hanging and seen from the rear. It was found in a cave at Le Manie, Finale Ligure, Italy.

Fig. 44 Zoomorphic sculpture of a lion's head. Found on the slopes of Monte Gazzo at Sestri Ponente, near Genoa, Italy. P. Gaietto Collection.

Fig. 45

Fig. 46

Fig. 45 Lioness with a human body. Ivory. Aurignacian, 32,000 years old. Hohlenstein Cave, Stadel, Germany. Ulm Museum.
Copyright 2008, Gaura, Wikimedia, public domain, copyright expired.

Fig. 46 Lioness carved on an ivory plate. Pavlovian. Moravia, Czech Republic.

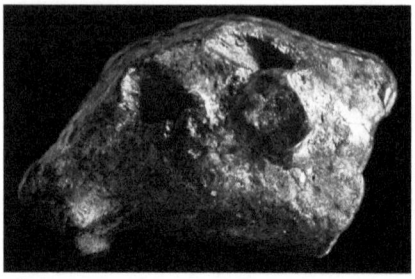
Fig. 47

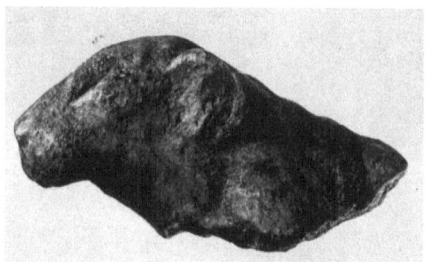
Fig. 48

Fig. 47 Sculpture of a lioness head found at Dolní Věstonice, Moravia, Czech Republic.
Fig. 48 Sculpture of a lioness head found at Dolní Věstonice, Moravia, Czech Republic.

7

Leopard or panther

The leopard or panther (*Panthera pardus*) (Fig. 49) is an animal of warm climates. During the Paleolithic it was widespread in central and southern Europe, and during the Neolithic period it still lived in Turkey.
In some areas the fossil record could be from the snow leopard and these may have been misinterpreted. The Snow Leopard *(Panthera uncia)* still lives today in the Himalayas and Central Asia as far as Manchuria; it is larger than the leopard.
In my attribution of Paleolithic sculptures I refer to a generic "panther" and this distinguishes it also from the lion. Today the leopard or panther lives in Africa and South Asia.
In Turkey, 32 miles from Konya is the site of Çatalhöyük, an ancient city which flourished between 6,800 and 5,700 BC. Images of a religious nature were found there including paintings of two leopards facing each other (Fig. 50) and, in a style similar to later South Asian religions, two small lion sculptures alongside a mother goddess on a throne, and some wall paintings of the great Bull, a parallel deity.
Çatalhöyük is among the oldest known cities, with an estimated population of 5,000, while the oldest temple with an altar for sacrifices and a panther sculpture was found in northern Spain at El Juyo in Santander province. The El Juyo temple consists of a cave in which an altar was built consisting of a rectangular monolith weighing over 2,000 pounds on which offerings of vegetables were placed. Opposite the altar was a bicephalic zoo-anthropomorphic sculpture (see Fig. 51). The American and Spanish paleoethnologists who discovered and studied it interpreted it as representing half a human face and half a panther. It is 12 inches in height. This sculpture is 14,000 years old (Upper Paleolithic) and has been attributed to the Magdalenian phase. My attribution regarding spiritual culture is Post-Aurignacian, since the Magdalenians had no two-faced divinities; they did not produce monoliths and they did not sculpt stone.
In the nearby cave of Altamira there is a painted giraffe, another warm-climate animal, which belongs to the Magdalenian artistic culture. It follows that the El Juyo panther, even if Post-Aurignacian, could be a leopard and not a snow leopard. For the purpose of representation we'll call it a panther.
The sculpture in Fig. 52 belongs to the Upper Paleolithic. It is 2.7 inches long and made of mammoth ivory. It is likely to be a cold-climate panther, i.e. a snow leopard *(Panthera uncia)*. This can be deduced from the volume of the body which is greater than that of a leopard.
In Fig. 53 the 3.5-inch sculpture was found in the same Vogelherd Cave in which one of the previous sculptures was found, but the body is more massive. Maybe it's a snow leopard *(Panthera uncia)*. Its body is decorated with incisions of dots and lines that, when crossed, form rhombuses. It's carved in mammoth ivory. The German paleoethnologists who discovered and studied these two ivory sculptures have labelled them generically as "felines".

 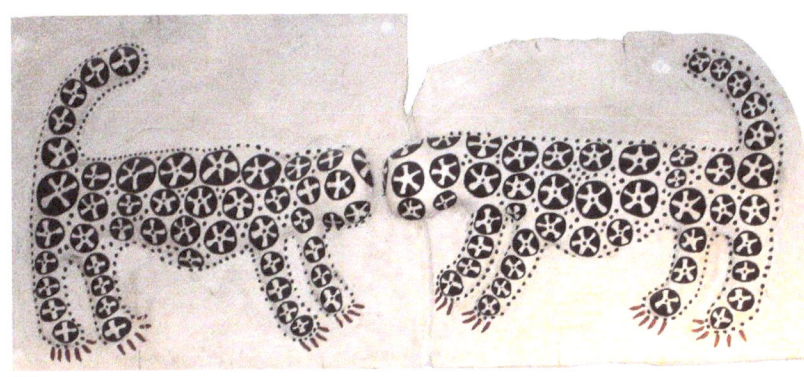

Fig. 49

Fig. 50

Fig. 49 Leopard or panther (*Panthera pardus*). Lives in Africa and in South Asia. Museum of Natural History "Giacomo Doria", Genoa, Italy

Fig. 50 Two leopards facing each other. A plaster mural painting. Extinct religion of the pre-Neolithic civilization of Çatalhöyük (present-day Turkey).

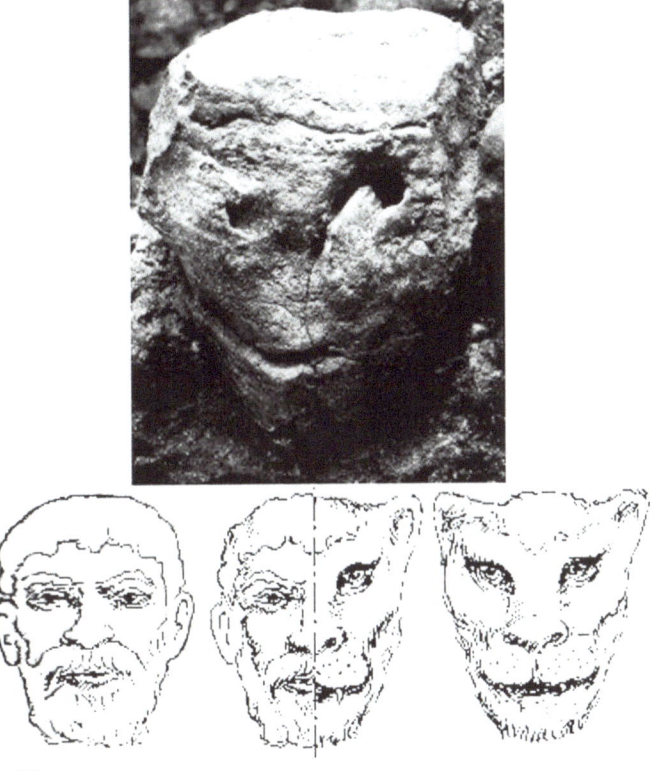

Fig. 51

Fig. 51 A two-faced zoo-anthropomorphic sculpture from the Paleolithic found at El Juyo, Santander, Spain. Altamira Museum, Santander.

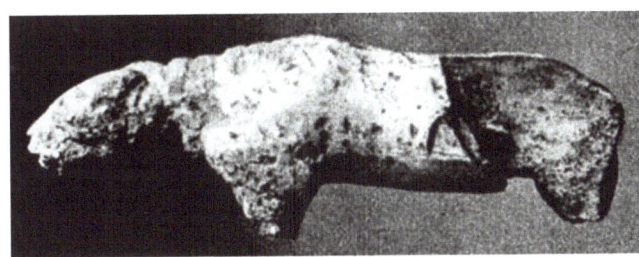 

Fig. 52

Fig. 53

Fig. 52 Panther or snow leopard *(Panthera uncia irbis)*. Recent interpretations consider this sculpture a lion. Vogelherd Cave, Germany.

Fig. 53 Panther or snow leopard *(Panthera uncia irbis)*. According to recent interpretations, it could be a lion. Vogelherd Cave, Germany.

8

Horse

The horse *(Equus caballus)* (Fig. 54) was domesticated around 5,000-10,000 years ago.

In the evolution of *Equidae*, the *Equus stenonis* of the early Pleistocene is linked to zebras by its type of dentition. *Equus hydruntinus*, which survived up until the Mousterian, shows remarkable affinities with *Equus stenonis* but it also possesses synthetic characters currently found distributed among other *Equids*: the donkey, horse and onager *(Equus hemionus)*, an Asiatic wild ass.

During the Paleolithic the *Equidae* were hunted by humans as food. In some caves of France there are many paintings of *Equidae* from the Upper Paleolithic (Magdalenian) dating to between 18,000 and 11,000 years ago. The reason why these were made has not been determined, and interpretations of paleoethnologists differ. The horses were not gods. Perhaps the images were painted to propitiate hunting or to celebrate the killing of horses for initiation rites.

These cave paintings allow us to subdivide the depicted horses into types. The most frequently found is Przewalskii's horse *(Equus caballus przewalskii)* which still lives in western Mongolia. The Celtic horse (Breton or Shetland Island poney) is less frequent and the Nordic horse and an uncertain donkey-onager are rarer.

I do not know of any horse among the gods from the historic period, though it was considered a sacred animal and many names of gods were preceded by the prefix *ippo-*. The horse was also present in some civilizations in spiritual culture, as in the Greek civilization where it was seen in the form of a centaur (Fig. 55), half man and half horse. In Fig. 56 we see a head that looks like a horse, without neck and with a mane. The muzzle is facing right. It is 10.2 inches long. I attribute it to the Mousterian, but it could be Aurignacian.

The sculpture in Fig. 57 is a decoration on a spear propeller from the Magdalenian made of deer antler. The theme realized, understood as a cycle of life, even human, is part of a deep spiritual culture. Below is the head of a very young horse and to the right is that of an adult specimen and, facing up, a horse skull. The subject of this work can be interpreted as birth, life and death, a philosophical theme that in my opinion goes beyond the idea of decoration. It is the only one of this type that I know of as far as the Paleolithic is concerned, but in all the proto-historic, primitive and historicl civilizations of the entire world, there are depictions of human and animal skulls, although not very frequent, which symbolize death. The sculpture in Fig. 58 is in mammoth ivory; it is two inches long and dates to 32,000 years ago. It was found in Germany in the same cave where the two panther sculptures were found (Figs. 52 and 53). It's from the Aurignacian civilization. Observe the similarity in shape of the head of this horse with the horse head of Fig. 56. This one, in ivory, shows details of the snout that the other does not possess, as previously these were not depicted.

Fig. 54

Fig. 55

Fig. 54 Head of a horse (*Equus caballus*).
Fig. 55 A centaur. Vatican Museums.
Copyright 2009 Wknight94, GNU Free Documentation License, Creative Commons Attribution-Share Alike 3.0 Unported.

Fig. 56

Fig. 56 Horse's head with mane (muzzle to the right). Urbe, locality of Vara (Savona), Italy. P. Gaietto Collection.

Fig. 57   Fig. 58

Fig. 57 Remains of a spear propeller from the Magdalenian, representing a horse head, foal and a horse skull. Found in the Mas d'Azil Cave in Ariège, France. National Archaeological Museum, Saint-Germain-en-Laye.
Fig. 58 A horse carved in mammoth ivory. Vogelherd Cave, Germany.

9

Moose

The modern moose *(Alces alces)* is a ruminant artiodactyl mammal of the *Cervidae* family (Fig. 59). Moose are larger than deer: they measure up to 10 feet in length and 6.5 feet in height at the withers. The antlers are only present in males. Moose inhabit the circumpolar area (north of Eurasia and North America). The American moose *(Alces americanus)* lives in North America.

The extinct Quaternary moose *(Alces latifrons)* had a more powerful structure than the present one. It lived throughout Europe and in the warm periods of the Quaternary withdrew north; in the cold ones it went south, like the Royal auk (*Alca impennis* or *Pinguinus impennis*), reaching the shores of the Mediterranean of Spain and southern Italy (Fig. 1). In the Paleolithic it was always hunted by humans for food.

To my knowledge, in the first historic civilizations the moose is not among the sacred animals, since these civilizations all arose in areas with a warm climate. But I would not exclude that in the North of Eurasia, among little-known populations, the moose was indeed a sacred animal, whether it was or was not depicted in sculpture, as the range of sacred animals is quite diverse and numerous (it is a sacred animal for Native Americans and is also a constellation formed by various stars in a sky chart the Lapps engraved on stone 4,000 years ago).

The seven sculptures I present here, depictions of moose heads, are based on interpretation of the shape of the snout, for the upper and lower profile of the snout is bent downwards. A stylistic deformation must be taken into account, which can be more or less accentuated in length. The antlers of these animals are not depicted, like the tusks of mammoths. These seven sculptures have an approximate date between 300,000 and 80,000 years ago, in fact the processing technique used for some is the Acheulean type, for others Clactonian, among those made of flint. The other sculptures in other types of rock were slightly damaged by rolling; we analyze their shape and style since traces of workmanship are rare. Style can permit a cultural attribution: the elongated style of these sculptures, for example, is not repeated during the Mousterian.

In Fig. 60 we see a moose head of the neckless type. It is a typology also present in sculptures depicting human heads. The sculpture is in flint and 6.6 inches long. The photo below shows how the area was excavated to highlight the jaw. Just like the mammoth head without a neck in Fig. 15, it was hollowed out underneath for the same purpose. Fig. 61 represents a moose head on a body without limbs. It is in flint, 2.5 inches long. Look at the drawings for the section of the head and of the body.

The bicephalic sculpture in Fig. 62 unites a human head (left) with one with mixed features (an artistic invention) representing a human along with an animal which could be a moose, but this is a hypothetical interpretation. It is in flint and 3.9 inches long. The sculpture in Fig. 63 is of considerable size, measuring 18.8 inches in length. It is bicephalic. On the left is an elk head in elongated style that has kept its shape well. It is a little damaged by rolling. The remaining traces of processing still highlight the mandible. The head on the right is human and only some facial features are removed; the absence of a forehead and chin is typical of *Homo erectus*. In Fig. 64 a moose head (on the left, like the previous one) is joined to another that could be a bison head, but it is impossible to interpret further due to alluvial damage. It is 3.5 inches long and was found on the ancient alluvial plains of the Scrivia Torrent near Tortona, on the northern side of the Ligurian Apennines in the Po Valley. Fig. 65 shows a sculpture from Aarhus in Denmark, in flint, four inches long, having the same shape and style as the sculpture in Fig. 64. The moose head is on the left, combined with one that looks like a bison's. In Fig. 66 a moose head is joined to a human's, clearly visible on the opposite side. The style is elongated. It is of flint and seven inches long. Note that among bicephalic sculptures of the Acheulean, one head is frequently more accurately executed than the other.

Fig. 59 Moose (from a children's cartoon). Observe the deformation of the muzzle, useful for interpreting the sculptures.

Fig. 59

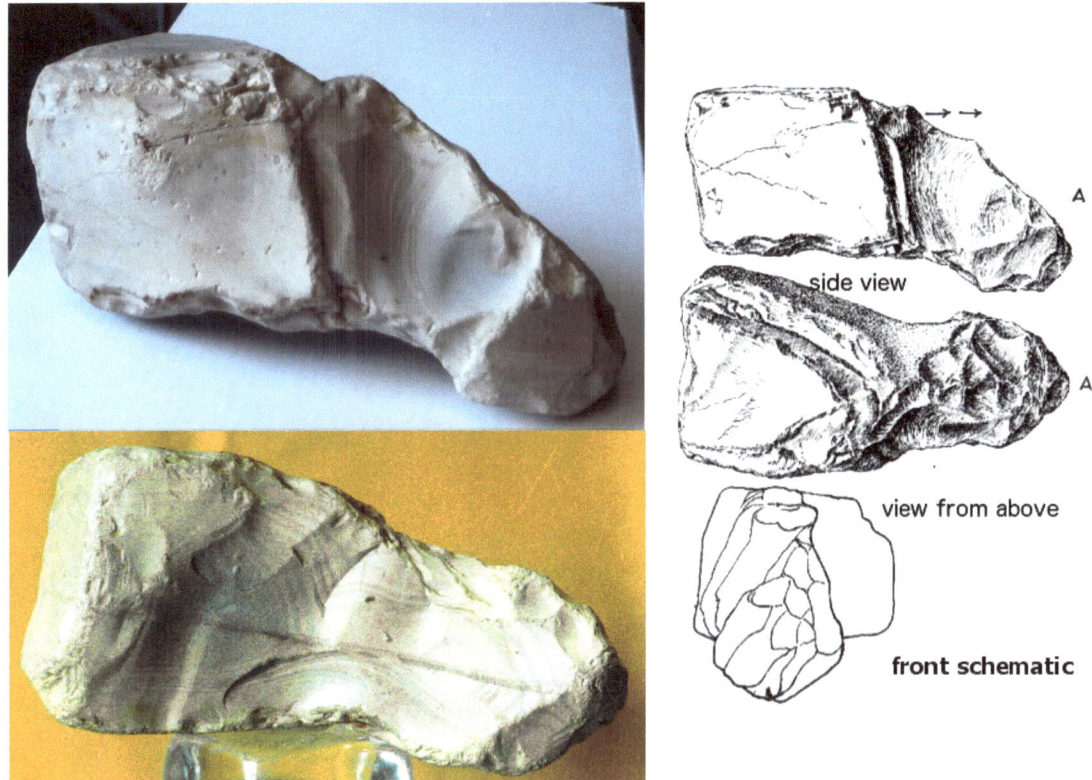

Fig. 60

Fig. 60 Moose head without neck. Flint. (The first photo, top left, is a side view. The drawings are side, top, front. The second photo below is a bottom view). Found at Rodi Garganico (Foggia), Italy. P. Gaietto Collection.

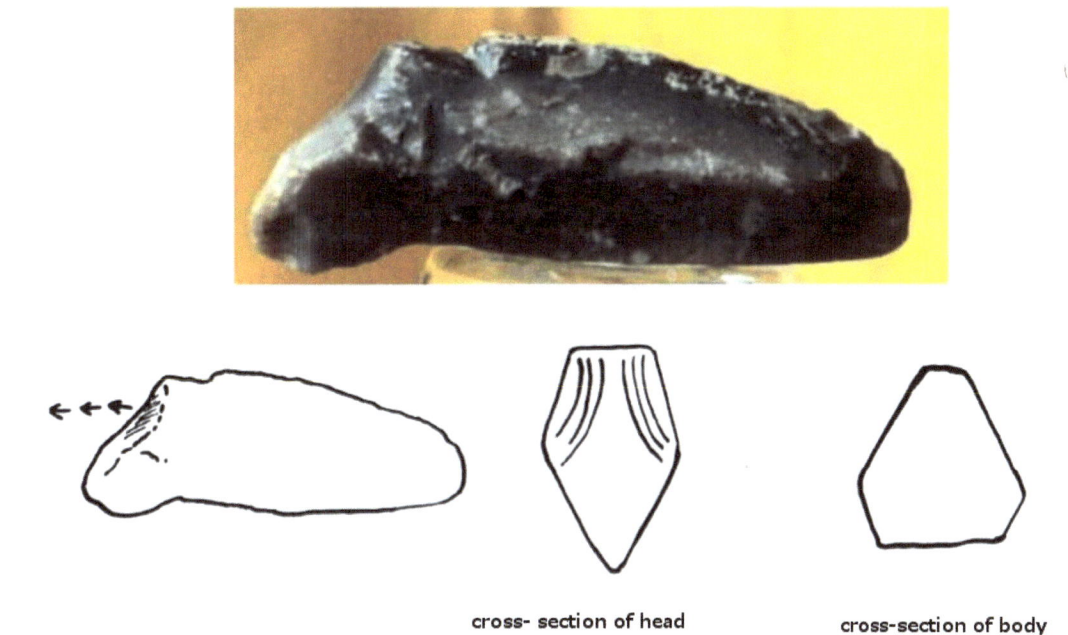

Fig. 61

Fig. 61 A sculpture representing a moose head with body and without limbs. Found at Rodi Garganico (Foggia), Italy. P. Gaietto Collection.

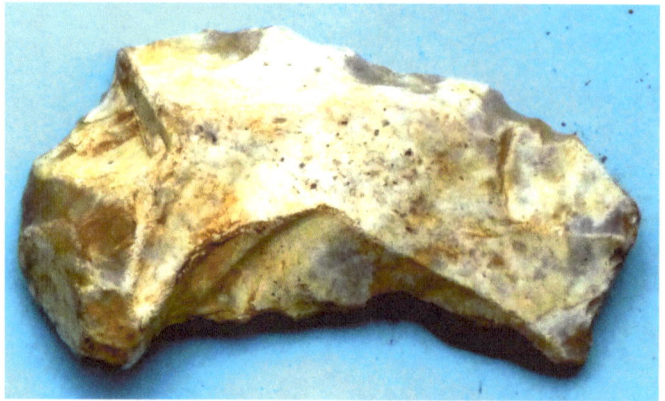

Fig. 62

Fig. 62 Sculpture representing a human head (left) joined to an artistic hybrid man-animal head which could be a moose. Found at Rodi Garganico (Foggia), Italy. P. Gaietto Collection..

Fig. 63

Fig. 63 Representation of a moose head (left) joined at the neck with a human head. The moose head is sculpted in an elongated horizontal style. Masone (Genoa), Italy. P. Gaietto Collection.

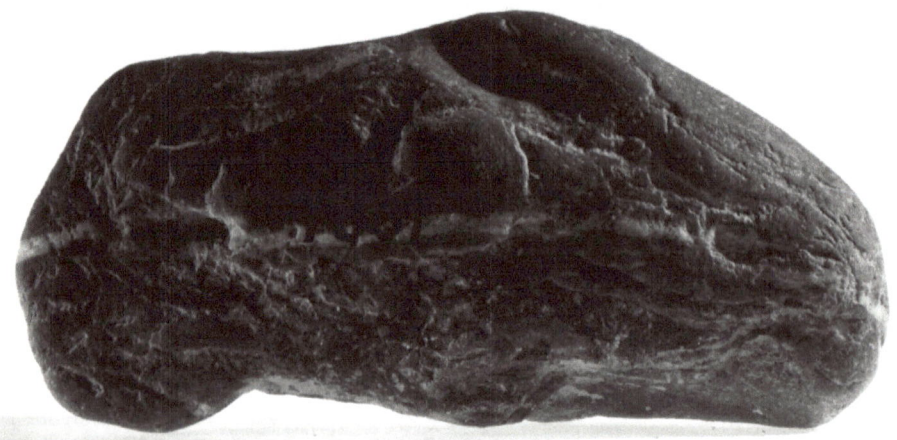

Fig. 64

Fig. 64 Representation of a moose head (left) joined to another animal head, perhaps a bison. Tortona (Alessandria), Italy. P. Gaietto Collection.

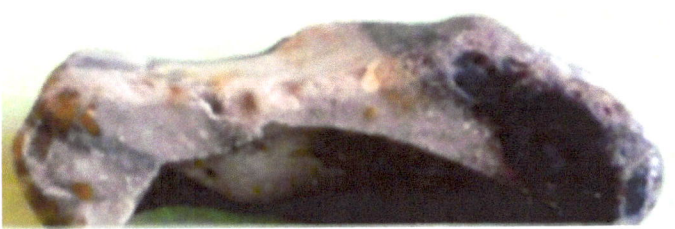
Fig. 65

Fig. 66

Fig. 65 Representation of a moose head (left) joined at the neck to another head, perhaps a bison's. Found at Aarhus, Denmark. P. Gaietto Collection.
Fig. 66 Representation of a moose head (left) joined to a human head without forehead and with prominent jaw. Flint. Found at Rodi Garganico (Foggia), Italy. P. Gaietto Collection.

10

Goat

Some types of goat and rams lived in Eurasia since the Pliocene, 2 million years ago, before the appearance of humans.
The ibex *(Capra ibex)*, now rare, has taken refuge in the mountains. In the cold periods of the Middle and Upper Paleolithic, up until the Magdalenian, it was widespread. Among the Caprids, the European mouflon *(Ovis musimon)* is now found only in Sardinia and Corsica. The antelope family, which has many living species both in Africa and Asia, was present during the Paleolithic with chamois and saiga antelope. The chamois *(Rupicapra europaea)* survived by taking refuge in the Alps, the Apennines and peaks of the Pyrenees. The saiga antelope *(Saiga tatarica)* spread from east to west, that is from China to France, and reached England. It currently roams the steppes of central Asia. The post-Paleolithic domestic goat *(Capra hircus)* is found throughout the world.
The Paleolithic sculptural representations I have identified as "goat" are all without horns and are of different types and artistic styles, so it is not possible to interpret them as mouflon, ibex, chamois or saiga; they could also be other animals.
The ram is the male goat (Fig. 66). In the extinct religion of ancient Egypt, the god Khnum had a human body and a ram's head (Fig. 68). The demon Naigamesha in the second century AD was represented with a human body and a ram's head; it belonged to the Hindu and Jain religions.
In Fig. 70 a goat's head is represented in a bicephalic Acheulean sculpture in an elongated artistic style with a bison head (perhaps) attached to the neck. Both animals are represented without horns. The sculpture is in flint and 7 inches long.
The bicephalic sculpture in Fig. 71 depicts a human head (left) joined with a goat's head. It is nine inches long. The type of rock is slightly damaged by erosion, but the sculpted shape has not been altered. Acheulean.
In the bicephalic sculpture of Fig. 72 we see two goat heads joined at the neck in an elongated artistic style. The sculpture is 9.8 inches long. The type of rock is slightly damaged by erosion, but the sculpted shape has not been altered. Acheulean. The bicephalic Mousterian sculpture in Fig.73 represents a human head (left) joined to a goat's head. It is 2.7 inches high

Fig. 67   Fig. 68   Fig. 69

Fig. 67 A male goat or ram. Museum of Natural History "Giacomo Doria", Genoa, Italy.
Fig. 68 The Egyptian god Khnum with a ram's head. External wall of the Temple of Latopolis Magna (today Esna), Egypt.
Copyright 2007 Merlin-UK, GNU Free Documentation License, Creative Commons Attribution - Share Alike 3.0 Unported.
Fig. 69 The divinity Naigamesha, a man with a ram's head. Hindu and Jain religions. Uttar Pradesh, India. Second century BC.

Fig. 70

Fig. 70 A representation of a goat's head (left) with a bison's head attached at the neck. Madrid, Spain. P. Gaietto Collection.

Fig. 71

Fig. 71 A representation of a human head (left) joined at the neck with a goat's head. Locality Vesima, Genoa. P. Gaietto Collection.

Fig. 72

Fig. 72 Two goat heads joined at the neck in an elongated artistic style. Tiglieto (Genoa), Italy. P. Gaietto Collection

Fig. 73

Fig. 73 Sculpture of a human head (left) joined with a goat's head. Vado Ligure (Savona), Italy. P. Gaietto Collection.

Bison

Bison belong to the *Bovidae* family; we find them in Eurasia starting from the Pliocene.

The most important Quaternary bison is the aurochs *(Bos taurus primigenius)*, which spread from Eurasia to North America. Their remains abound throughout France and depictions are also frequent in Magdalenian cave paintings.

Since the end of the Paleolithic, around 11,000 years ago, the European bison *(Bison bonasus)*, a Eurasian species of bison (Fig.74) was present. In the Middle Ages the European bison was widespread in forests throughout Europe, but by 1880 only 600 were left. Today it survives in captivity and in modest numbers in the wild. The decline seems to be caused by a loss of reproductive abilities of females.

The cave paintings of bison in the Magdalenian are not to be considered divinities, but certainly they had a religious function in rituals linked to propitiation of hunting. This is one of the most shared hypotheses.

This animal is not present in sculpture as a divinity with a human body in the first historic urban civilizations of the Nile Valley, the Euphrates and the Indus.

The three bison sculptures I present here are from three different regions (Denmark, France, Italy) distant in space and also in time (Lower, Middle and Upper Paleolithic).

In Fig. 75, a sculpture of a bison with head (right) and body without limbs. It is in flint and 2.9 inches long. Acheulean.

The sculpture in Fig. 76 is bicephalic. It represents a bison head (left) joined at the neck to a human head. It is in flint. Length: 4.3 inches. Mousterian.

The sculpture in Fig. 77 is also bicephalic. It represents a bison head (on the right) joined to a human's. It is 14.9 inches high. Aurignacian.

Fig. 74

Fig. 75

Fig. 74 The American bison *(Bison bison)*. This species is similar to the European bison *(Bison bonasus)*. Yellowstone National Park, USA.
Copyright 2006 Joel McLendon, Creative Commons Attribution - Share Alike 2.5 Generic license.

Fig. 75 A sculpture of a bison with head (right) and body without limbs. Found at Roskilde Fjord (Denmark). P. Gaietto Collection.

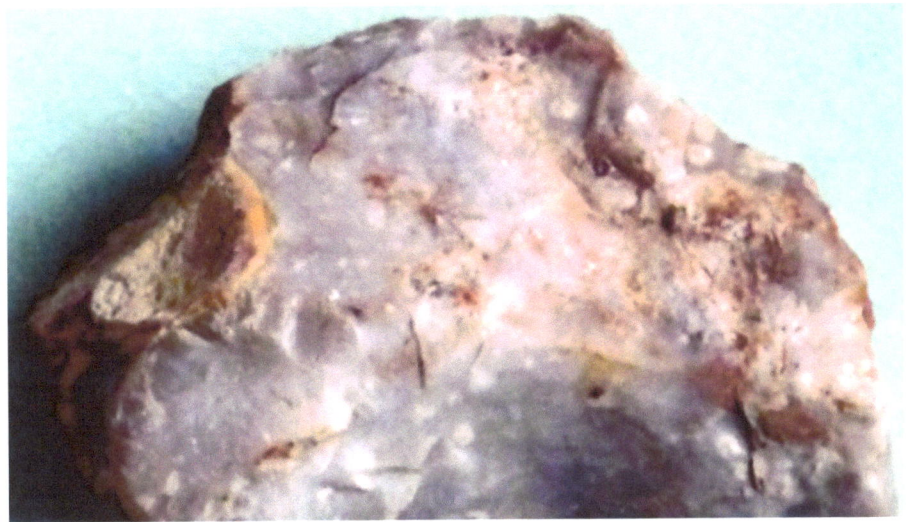

Fig. 76

Fig. 76 Sculpture of a bison head (left) joined at the neck to a human head. Found at Bonny-sur-Loire (Orléans, Loiret), France. P. Gaietto Collection.

Fig. 77

Fig. 77 Bicephalic sculpture of a human head (left) joined at the neck to a bison head. Masone, Genoa, Italy. P. Gaietto Collection

12

Bull

The domestic bull *(Bos taurus)* is found worldwide. The male is called a bull; the female, a cow. Its ancestor is the aurochs *(Bos taurus primigenius)*, both Afro-Asian and European, which we find beginning in the Pleistocene, and which in turn descended from *Bos planifrons*, widespread both in Asia and Europe. *Bos taurus primigenius* was present throughout Europe and northern Africa; larger than the ox, it lived throughout the Quaternary, disappearing in the Middle Ages. The last specimen was shot in Poland in 1627.

Both as a bull and as a cow (without horns), it is commonly found among the animals painted on cave walls by the Magdalenians, often shown racing and even in jumping positions. This behavior is not uncommon: oxen introduced by the British into wild Australia some centuries ago jumped 40 feet when racing. However, the bulls and cows painted in caves by the Magdalenians are not divinities; I hypothesize that they are linked to rituals of preparation for hunting, which were always religious practices. This is not surprising; in the first urban civilizations, religions were all polytheistic. It must be borne in mind, however, that each religion confers a different meaning on its own divinities, even if the gods had the same appearance – that is, human body and head of animal.

In sculptures found in the temple of Çatalhöyük, in addition to the bull, there were the mother goddess and the leopards that face each other, painted on plaster or as effigies painted in relief on the wall. The main deity was probably the bull, represented with only its head, modeled with earth and then dried. In some cases more than one bull's head was placed in the temples (Fig. 78). The bull was also painted in the temples in its complete body. The god Montu (Fig. 79) belonging to the polytheistic religion of Egyptian civilization, had a bull's head and a human body and was one of many gods. One of the deities of Hinduism is the god Nandikeshvara, a theriomorphic aspect of Shiva with a bull's head and a human body (Fig. 80).

Among the Paleolithic sculptures that I present it is difficult to establish which have been made simultaneously by the same people. Instead, it is possible through typology to assign them to the same period which often covers tens of millennia. In Fig. 81, a flint sculpture, 4.3 inches long, depicting a bull's head (left) joined to a human's. Acheulean. The sculpture in Fig. 82 is in flint and 3.5 inches long. It represents a bull's head joined to another of the same animal. Mousterian.

Fig.78

Fig. 78 Terracotta bull heads representing a deity of Çatalhöyük, from the first half of the sixth millennium BC. Museum of Anatolian Civilizations, Ankara, Turkey.

Fig. 79

Fig. 80

Fig. 79 A representation of the Egyptian god Montu with bull's head and human body from the temple of Montu, site of Medamud, Egypt. Limestone. Ptolemaic Period (332-30 BC). Louvre Museum, Paris.
Copyright 2010 Janmad, GNU Free Documentation License, Creative Commons Attribution - Share Alike 3.0 Unported.

Fig. 80 Statue of the god Nandikeshvara with a bull's head (Nandi) and a human body. Hindu religion (1,050-1,075 BC). Vamana Temple, Khajuraho, India.
Copyright 2012 Rajenver, GNU Free Documentation License, Creative Commons Attribution - Share Alike 3.0 Unported.

Fig. 81

Fig. 81 Sculpture of a bull's head (left) joined to a human head. Found at Aarhus, Denmark. P. Gaietto Collection.

Fig. 82

Fig. 82 Another sculpture of a bull's head joined at the neck to another bull's head. Found at Aarhus, Denmark. P. Gaietto Collection.

13

Bear

The cave bear *(Ursus spelaeus)* was of gigantic proportions. Standing on its hind legs, a male specimen could stand over ten feet tall and weigh 2,300 pounds. It lived throughout central Europe from the beginning of the Quaternary up to the Magdalenian and was essentially a cave species. In the Grotta della Bàsura cave at Toirano in Savona, it left imprints of its footsteps and claw marks on the walls, as well as a large number of skeletal remains.

Since the Magdalenian the brown bear *(Ursus arctos)* has replaced the cave bear. It still lives in the mountainous regions of Europe and America.

Humans have always hunted bears for food. In certain areas this animal was the subject of religious rituals. The oldest sacred places (of worship) of the Paleolithic have been discovered in central Europe and are related to bears. These were first found in 1917 in Switzerland (canton of St. Gallen) in the Drachenloch ("dragon's cave") by the Swiss Emil Bächler and his nine-year-old son, Toni. Excavations were conducted from 1917 to 1923 by Bächler, who found more than 30,000 remains of cave bears and in particular a bear skull with a femur attached to the cheekbone, which could only be the work of humans, and therefore an indication of a Paleolithic bear cult. Bächler identified a "place of sacrifice" with bear skulls, attributed to the Middle Paleolithic Mousterian civilization, the work of humans of the extinct Neanderthal species. About 20 inches from the walls were some manmade stone walls up to 32 inches high, with "rooms" between the walls filled with skulls and other bones of cave bears. It was perhaps a ritual site to propitiate hunts. In a third cave six "sarcophagi" were found made of stone slabs and containing numerous bear skulls laid out in neat arrangements. Bear skulls were also found placed in niches in rock, and another skull was found covered with small stones.

The same cult rituals were also practiced in Germany. In the Petershohle cave near Velden (Franconia), numerous niches were carved into the wall and filled with cave bear skulls.

Signs of the same rituals were also found in the Drachenhohle (dragon cave) near Mixnitz (Northern Styria, Austria) where, in addition to the remains of open hearths and stone tools (65,000-31,000 BC), the oldest traces of humans in Austria, a pit was found containing 30 cave bear skulls and some long bones of bears.

These rituals of worship of Neanderthals can be observed in modern Arctic civilizations and consist of supplying skulls and bones of bears to a superior being on whom the success or failure of hunting depends. It is therefore a religious practice.

The sculpture in Fig. 83 is in flint and 3.1 inches long. It represents a bear's head with open mouth, probably in an aggressive attitude, joined at the neck to another bear's head. Mousterian. In Fig. 84 is another sculpture like the previous one representing a bear's head joined at the neck to another bear's head. It is in flint and 2.6 inches long. The sculpture in Fig. 85 depicts a human head (left) joined at the neck to a bear's. It is in flint and 3.1 inches long. In Fig. 86 is a sculpture of a human head (left) joined at the neck to the head of a bear. It is 7.4 inches long, Mousterian. The sculpture in Fig. 87 is very small, just 1.1 inches long. The person who discovered it interpreted it as a "stylized animal". My interpretation is: bear head complete with body but without limbs. The sculpture of Fig 88 is 1.5 inches high; it represents a bear's head. It was found in the cave called Grotta della Bàsura at Toirano in Liguria, inhabited for long periods by cave bears. A "bear cemetery" was found in the cave. A human head and upper body (left) joined to a bear's head is shown in the sculpture in Fig. 89, 1.5 inches high.

Fig. 83

Fig. 84

Fig. 83 Bicephalic sculpture of a bear's head with wide-open mouth, joined at the neck with another bear's head. Roskilde Fjord, Denmark. P. Gaietto Collection.

Fig. 84 Bicephalic sculpture of a bear's head joined at the neck with another bear's head. Senigallia (Ancona), Italy. P. Gaietto Collection.

Fig. 85   Fig. 86

Fig. 85 Sculpture of a human head (left) joined at the neck to the head of a bear. Senigallia (Ancona), Italy. P. Gaietto Collection.

Fig. 86 Representation of a human head (left) joined at the neck to the head of a bear. Locality Vesima, Genoa, Italy. P. Gaietto Collection.

Fig. 87   Fig. 88

Fig. 87 Sculture of a bear's head complete with limbless body. Found at Kostenki, Russia. Photo by A. N. Rogachev.

Fig. 88 A bear's head sculpture found in Grotta della Bàsura cave, Toirano (Savona), Italy. P. Gaietto Collection.

Fig. 89

Fig. 89 Sculpture of a human head and body (left) joined to a bear's head. Locality of Palo, Urbe (Savona), Italy. P. Gaietto Collection.

14

Dog

The domestic dog *(Canis lupus familiaris)* (Fig. 90) appears at the beginning of the Neolithic, but we know nothing about wild Paleolithic dogs. The modern dog is a particularly polymorphic "species" which probably had various origins in different parts of the world, from different species of wild dogs that humans selected for different uses. From this we can also guess the major differences in shape and size between one dog "breed" and another.

The dog was a sacred animal for some historic civilizations and was even personified in the form of a god. Anubis was an Egyptian god with a dog's head and a human body (Fig. 91). Another dog-headed deity with a human body is Amida (Fig. 92), incarnation of the Buddha Amitābha, representative of a religious confession that arose in Japan. These historic civilizations already possessed the domestic dog, while the Paleolithic civilizations knew only the wild dog; but I hypothesize that in some religions there was an evolutionary link between the wild dog and the domestic dog, that was a sacred animal.

It is to be noted here that while technology is rapidly evolving, religions are slow to evolve; in fact, the main religions during the last 1,000 years have remained essentially unchanged, while technology has evolved continuously and with increasing intensity. Also, the passage from representing the head of a sacred animal in flint during the Paleolithic to the sculpting in marble of the head of the same species with a human body is mainly due to technological progress.

The sculpture in Fig. 93 depicts a human head (right) joined to a dog's head. It is 7.8 inches long, from the Acheulean.

In Fig. 94 a dog's head (on the right) is represented joined to a human one. Height 2.7 inches. Flint. Acheulean.

The sculpture in Fig. 95 depicts a human head (right) joined to a dog's. Flint. The length is three inches. Mousterian.

In Fig. 96 we see a dog's head (on the right) joined to a human one in a geometric and symbolic style, even if at first glance it does not seem a human head. The dog has a round eye worked with clear intent, like the entire contour of the sculpture. However, it is an atypical work. It is in flint, 5.9 inches high. The cultural phase could be Mousterian, but it must be kept in mind that in every age there have been populations with different skill in sculpting, both in general and at the individual level.

The sculpture in Fig. 97 depicts a dog's head viewed from above. It has large ears and a long and narrow snout on both sides. It was hollowed out below to highlight the shape of the jaw. It is 1.6 inches long.

Fig. 90

Fig. 90 The domestic dog *(Canis familiaris)*.

Fig. 91

Fig. 92

Fig. 91 The Egyptian god Anubis with dog's head and human body. The divinity is always sculpted. This drawing is modern, probably of a painting found inside a pyramid.
Copyright 2012 Perhelion, GNU Free Documentation License, Creative Commons Attribution - Share Alike 2.5 Generic.

Fig. 92 The god Amida with a dog's head and human body. A drawing from 1808 by a European artist of a sculpture in a Japanese temple.

Fig.93

Fig. 93 Bicephalic sculpture of a human head (right) joined at the neck to a dog's head. Found at Palo Locality, Urbe (Savona), Italy. P. Gaietto Collection.

Fig. 94

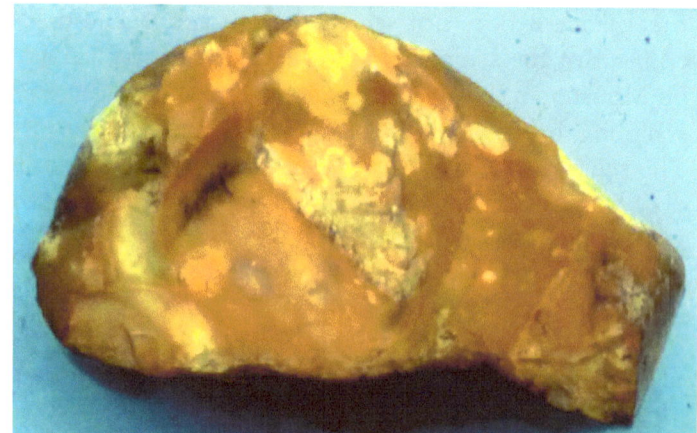
Fig. 95

Fig. 94 Bicephalic sculpture of a dog's head (right) joined at the neck to a human head. Found at Pescara, Italy. P. Gaietto Collection.

Fig. 95 Bicephalic sculpture of a human head (left) joined at the neck to a dog's head. From Peschici (Foggia), Italy. P. Gaietto Collection.

Fig. 96

Fig. 97

Fig. 96 Bicephalic sculpture of a dog's head (to the left) joined at the neck to a human head. From Rodi Garganico (Foggia), Italy. P. Gaietto Collection.
Fig. 97 Sculpture of a dog's head found at Tiglieto (Genoa), Italy. P. Gaietto Collection.

15

Seal

It is probable that the harbor seal *(Phoca vitulina)* (Fig. 98) or an earlier species of seal that lived during the Quaternary period arrived on Mediterranean coasts in colder periods.
The Quaternary seals, acclimatized to periods of warmer climate, are still present on Mediterranean beaches. The one species still present is the monk seal, which lives in temperate zones of the Atlantic and on the coasts of Sardinia.
I have no knowledge of depictions of seals in paintings from the Paleolithic: in fact, artistic finds are almost exclusively from continental areas where there are no seals.
Up until the last century, wooden masks resembling the head of a seal were used by the Eskimos (Inuit) of North America in ritual ceremonies of their religion. They are probably still produced. The Eskimos, according to various anthropologists of the 20th century, originated from the Magdalenian civilization which produced cave paintings of animals and which, in the Upper Paleolithic, migrated north to avoid climatic changes.
The fossil record of Magdalenian humans is from the Chancelade type, a Mongoloid similar to modern-day Eskimos. The religion depicted in the art of Eskimos is similar to that of the Magdalenians.
The sirens of Greek mythology, present in sculpture and in popular imagination as women with legs replaced by the body of a fish (Fig. 99), displayed the habits and physical forms of seals which in ancient times populated the beaches of many Greek islands.
I hypothesize that the sculpture in Fig. 100, which I previously described as a mammal's head with a neck, represents instead a seal head with a neck. It comes from Liguria, a mountainous region along the sea, where the seal was present. Being an animal that lives in the sea and on land, endowed with its own beauty, the seal could have been a sacred animal.

Fig. 98   Fig. 99

Fig. 98 The crabeater seal *(Lobodon carcinophagus)* inhabits the seas of the circumpolar region.
Copyright 2008 Finavon, CC Public Domanin, National Oceanic and Atmospheric Administration (National Marine Mammal Laboratory).
Fig. 99 Statue of the Little Mermaid. Bronze, 1913. By E. Eriksen. Port entrance at Copenhagen, Denmark.
CC BY 4.0 © 2014 Jose Antonio

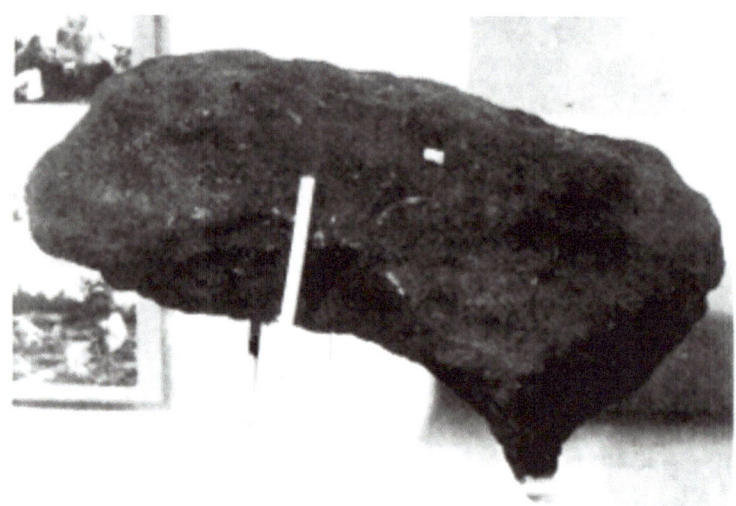

Fig. 100

Fig. 100 Sculpture of a seal's head and neck found at Urbe (Savona), Italy. P. Gaietto Collection.

16

Bird

The Quaternary birds are little known since it is particularly difficult to define them from bones found on archaeological sites which are generally leftovers from human meals. In terms of their relationship to the large mammals for which these difficulties do not exist, the birds of the Quaternary were similar to modern ones.
The birds worshipped by humans as "sacred animals" were the great flyers and migrators such as the gull (Fig. 101), the flamingo (Fig. 102), the ibis and the swan; and birds of prey such as the eagle, hawk (Fig. 103), vulture (Fig. 104) and others.
In historic periods, as already seen with mammals, sacred birds were depicted as gods with human bodies. The

Assyrian bas-relief in Fig. 105 depicts an eagle's head with a human body and wings and represents a genie. Ninth century BC. In the painting in Fig. 106 we see the Egyptian god Horus with a falcon's head and a human body. The bas-relief in Fig. 107 represents another Egyptian god, Thoth, with an ibis head and human body.

In the Assyro-Babylonian, Phoenician, Greek, Roman and Christian religions, the winged man could be from time to time a god, a demon, a genie, an angel. In the Christian religion the angel with wings (Fig. 108) is a "pure spirit created by God as His messenger to humanity".

I have not interpreted the birds depicted in Paleolithic sculpture as species, this being impossible, and I have simply called them birds, since they have a head and a beak. Likewise, scholars of Magdalenian cave paintings failed in 1900 to interpret any species of painted birds in the caves.

The only zoomorphic tool I present is a biface. The biface is a typical tool of the Lower Paleolithic. The older types have a coarse bill, while more recent ones are refined, like this one that belongs to the Final Acheulean. Bifaces are tools that are gripped, resting on the palm of the hand. Of varying sizes, from 2.3 to 9.8 inches long, the heaviest exceed 1,500 grams in weight while the average values are 4.7 inches and 250 grams.

The paleoethnologists of 100 years ago recognized only the authenticity of the stone tools and did not perceive the anthropomorphic and zoomorphic sculpture of the Lower and Middle Paleolithic, which they called random "stone figures". However, they had considered the biface a tool "more beautiful than pure necessity", that is they recognized the origin of the embellishment.

In the first historic civilizations two forms of art were always present:
1) figurative art, representing a divinity or idea;
2) decorative art that embellished objects.

The bifaces have two types of embellishment, similar to the objects of post-Paleolithic civilizations: 1) the common biface has a "harmonious and symmetrical" form; 2) the zoomorphic biface has a "figurative" decoration which mimics the head shape of the swan or other bird.

The biface shown has the shape (in side view) of the head of a swan (Figs. 109 and 110). The bird's head is the decoration of the tool. This sculpted swan head is exactly the same as a schematic swan-head design without a neck. It is a functional tool and has never been used. I consider it a ritual object due to its beauty and rarity.

In protohistory the swan was a sacred animal among northern peoples. In Greek-Roman civilization the religious myth of Leda and the Swan was popular. In fact, numerous sculptures of the sexual union between the swan and young Leda are found in Europe, Asia and Africa, in the territories of the Roman Empire.

The Lower Paleolithic sculpture in Fig. 111 depicts a two-faced head from the early Acheulean, consisting of a bird's head with a large beak (left) joined at the neck to a human head. Observe the eye of the human head obtained with a single blow on the hard flint. Length: 4.1 inches. The sculpture in Fig. 112 is the back of the one in Fig.111, but it seems to be another sculpture. It must be considered that this work is about 300,000 years older than the zoomorphic biface (Figs. 109 and 110) and the processing technique was coarser. The sculpture in Fig. 113 depicts a human head joined to a bird's (right). The bird, as in the previous sculpture, has a large beak and seems to be more important than the human head which has a geometric style. In Fig. 114 the sculpture represents a human head (on the left) joined to another head that looks like a bird's. On the back and partially below are the carved parts, while in the picture you can see the natural part that was used. The work is disfigured by alluvial rolling. It is Mousterian. The sculpture in Fig. 115 depicts a human head (left) joined to a bird's with a large curved beak. Acheulean. In Fig. 116 is a bird's head with beak (right). Length: 3.1 inches. Mousterian. The sculpture in Fig. 117 represents a head that seems a bird's, joined at the neck to a human head (to the right). The one in Fig. 118 depicts a bird's head (left) joined to a human's. The bird's beak has part of the natural shape of the flint nodule, but the processing is evident. It has rolling damage. It can be considered an atypical sculpture. The largest sculpture I present is that of Fig. 119. It represents a bird's head with a large beak (on the left) joined at the neck to a human head. Final Acheulean. The sculpture in Fig. 120 represents a human head (on the left) joined to one of a bird. Length: 11 inches. It was found in Tiglieto (Genoa), at the same location as the previous one. The bird's beak resembles the profile of the face of a human head in an elongated style but, if turned, it can be interpreted as a beak. The sculpture in Fig. 124 depicts a human head (right) joined to a bird's with a large beak, which is also realistic. Mousterian. The sculpture in Fig.122 comes from Greece and is of a type that seems not to have been found in Western Europe. The bird is depicted with the entire head, neck and part of the body (above) joined to a head that looks human. The work is quite disfigured, but I decided to include it because it can be a clue and a stimulus for research.

Fig. 101   Fig. 102   Fig. 103   Fig. 104

Fig. 101 Seagull (*Larus fuscus*).
Copyright 2011 Magnus Manske, Creative Commons Wikimedia, public domain.
Fig. 102 Flamingo (*Phoenicopterus ruber*). Museum of Natural History "Giacomo Doria", Genoa, Italy.
Fig. 103 Falcon. Museum of Natural History "Giacomo Doria", Genoa, Italy.
Fig. 104 Collared Vulture. Museum of Natural History "Giacomo Doria", Genoa, Italy.

Fig. 105   Fig. 106   Fig. 107

Fig. 105 An Assyrian deity having an eagle head and human body with wings. Kalakh (Nimrud). From an extinct religion.
Fig. 106 An Egyptian divinity: the god Horus with a falcon's head and human body. The deity is always portrayed in sculpture. This is a modern drawing, probably a copy of a fresco found inside a pyramid.
Fig. 107 An Egyptian divinity: the god Thoth with ibis head and human body. This bas-relief is on back of the throne of the seated statue of Ramsese II in Luxor Temple, Egypt.
Copyright 2009 Jon Bodsworth, Wikimedia Commons from Egypt Archive.

Fig. 108

Fig. 108 A Christian religious statue. Angel (winged man). Castel Sant'Angelo, Rome. Sculpted by Ercole Ferrata. 1667.
Copyright 2007 Tetraktys, GNU Free Documentation License, Creative Commons Attribution - Share Alike 3.0 Unported.

Fig. 109

Fig. 110

Fig. 109 A zoomorphic biface, a ritual tool. It represents a swan's head (or similar bird) without neck. Side view. 4.9 inches long. Flint. Found at Vieste (Foggia), Italy. P. Gaietto Collection.
Fig. 110 Top view of biface in Fig. 109.

47

Fig. 111   Fig. 112

Fig. 111 Bicephalic sculpture of a bird's head (left) joined to a human head. Length: 4.1 inches. Rodi Garganico (Foggia), Italy. P. Gaietto Collection.
Fig. 112 Bird's head (right) joined to a human head. (Rear view of the sculpture in Fig. 111).

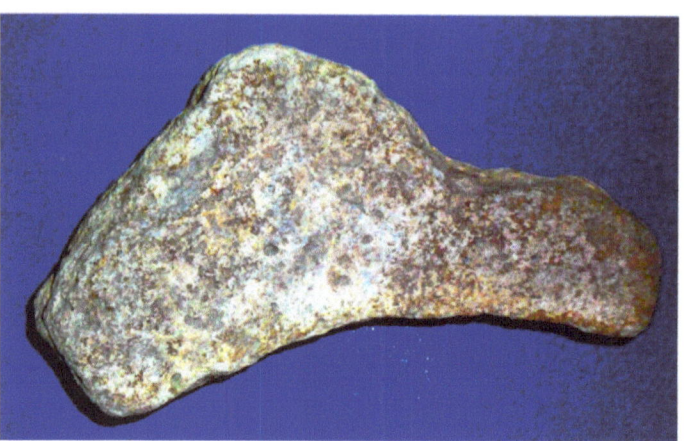 

Fig. 113   Fig. 114

Fig. 113 Sculpture of human head (left) joined to a bird's head. Length: 6.8 inches. Found at Karaman, Turkey. P. Gaietto Collection.
Fig. 114 Sculpture of human head (left) joined to a head that seems a bird's. Length: 5.5 inches. Monti Lessini, Pian Castagné (Verona), Italy. P. Gaietto Collection.

 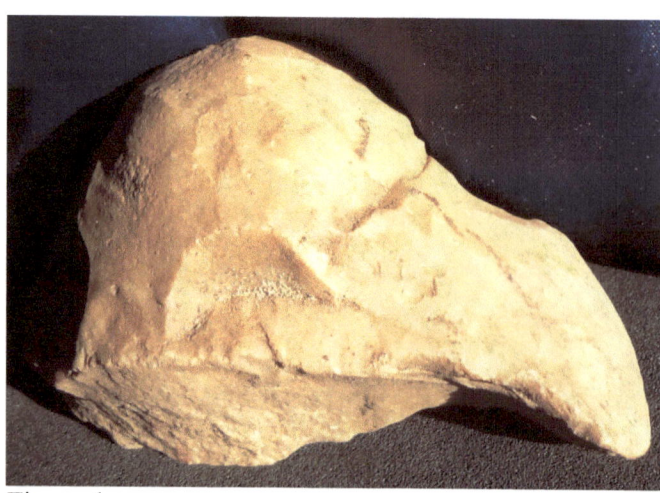

Fig. 115   Fig. 116

Fig. 115 Sculpture of a human head (on left) joined at the neck to a bird's head with a curved beak. Length: 3.3 inches. Flint. Rodi Garganico (Foggia), Italy. P. Gaietto Collection.

Fig. 116 Sculpture of a bird's head. Length: 3.1 inches. Périgueux, France. P. Gaietto Collection.

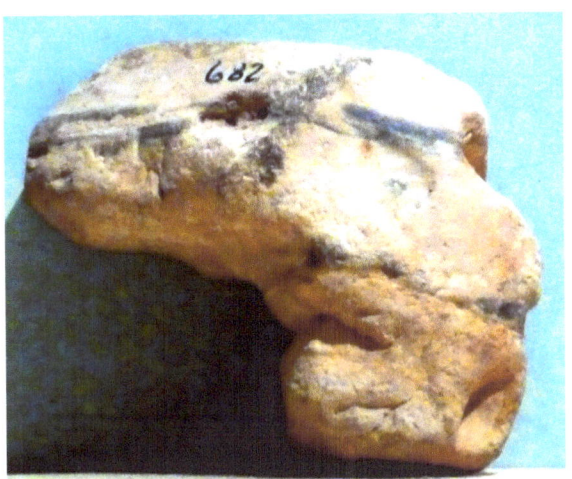

Fig. 117

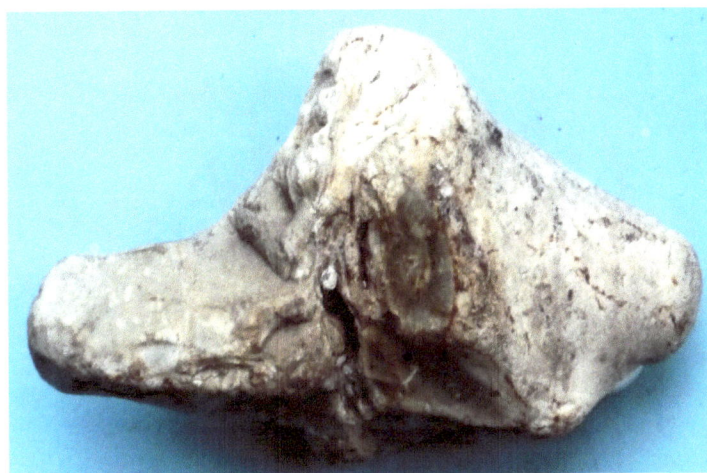

Fig. 118

Fig. 117 Sculpture of a bird's head (left) joined to a human head. Height: 2.7 inches. Flint. Rodi Garganico (Foggia), Italy. P. Gaietto Collection.

Fig. 118 Sculpture of a bird's head (left) joined at the neck to a human head. Length: 6 inches. Flint. Rodi Garganico (Foggia), Italy. P. Gaietto Collection.

Fig. 119

Fig. 120

Fig. 119 Sculpture of a bird's head (left) joined at the neck to a human head. Length: 13 inches. Tiglieto, Genoa, Italy. P. Gaietto Collection.

Fig. 120 Sculpture of a human head (right) joined to a bird's head. Length: 11 inches. Tiglieto, Genoa, Italy. P. Gaietto Collection.

Fig. 121   Fig. 122

Fig. 121 Bicephalic sculpture of a human head (right) joined to a bird's head. Length: 11.8 inches. Vesima Locality, Genoa, Italy. P. Gaietto Collection.
Fig. 122 Bicephalic sculpture of a bird's head with neck and part of the body (above) joined to a human head. Larissa, Greece. P. Gaietto Collection.

17

Fish

The fish or some species of fish are among the sacred animals. Like the snake, the fish is connected to the water cult, source of life, still practiced today in many cultures.
A "fish" god depicted in sculpture was found in various specimens at the site of Lepenski Vir, in eastern Serbia, in the center of the Balkan peninsula. The term "man-fish god" was used by archaeologists who discovered it, because it has a mouth with large lips that suggest the mouth of a fish.
The sculpture in Fig. 123 depicts a half-length man with his hands on his chest and his head looking upwards. Eyes, nose, mouth, hands and every other feature show a style that deforms the real, but increases the expression. This type of sculpture was produced by a Danube fishing people, and because of the shape of the lips it was called "Fish Man" and considered a god of fishing. The ascertained material culture is Mesolithic.
The Mesolithic lithic industries have a certain universality since they are micro-industries in flint and exhibit more or less geometric shape. Art, on the other hand, being a function of religion among different peoples with different traditions, was very varied.
In Europe, many types of Mesolithic art are known. I mention only three: 1) the anthropomorphic and zoomorphic sculptures of the Lepenski Vir type; 2) the colored pebbles painted with abstract images, identified in France in Mas d'Azil Cave and found at numerous European sites; and 3) the paintings in rock shelters in the Spanish Levant (Spain's Mediterranean coast) that depict hunting scenes, running warriors, fights and dances.
The sculpture from Lepenski Vir belongs to one of many lines in which stone sculpture evolved during the Lower, Middle and Upper Paleolithic. The pebbles with abstract paintings from Mas d'Azil could be connected to abstract incisions from the Middle and Upper Paleolithic; the paintings in the Spanish Levant derive instead from zoomorphic paintings of the Upper Paleolithic (Magdalenian) to which the representation of humans, complete with clothes, was added.

The Mesolithic separates the Paleolithic from the Neolithic and in certain territories of Europe and the Middle East it was very short or didn't even exist. In the Mesolithic period great displacements of populations and therefore of civilizations took place. The Mesolithic "man-fish" god (Fig. 123) in its style of representation looks like a Neolithic work.

A Roman sculpture (Fig. 124) found in France depicts three deities, one of which is three-headed with large lips in the shape of a fish's mouth, and which is usually interpreted as a god and a fish-man.

The three-headed divinity is an evolution in the depiction in sculpture of the two-headed or two-headed god, both anthropomorphic and zoomorphic and zoo-anthropomorphic. The Celts and Gauls had two-headed and three-headed divinities. In Bulgaria the three-headed Thracian knight-god was widespread.

The fish lips of the sacred being constitute the part that symbolizes the animal, like the wings of birds in angels. It can be extrapolated that every human head carved with large lips constitutes the "man-fish" deity.

The Paleolithic sculpture of Fig. 125 depicts a human head with large lips. This type of sculpture is at the origin of the deities I mentioned, who represent a man-fish in post-Paleolithic times. It is a work somewhat damaged by atmospheric events, but you can still see the engravings that adorned it and that were probably a hairstyle.

The bicephalic sculpture in Fig. 126 represents a "fish-man" head joined to a human one. The representation is completed by engravings partially obliterated by rolling. I believe it belongs to the Upper Paleolithic but I would not exclude the Mesolithic.

Fig. 123

Fig. 124

Fig. 123 "Fish Man" divinity with large lips. Mesolithic. Lepenski Vir, Serbia.
Copyright 2006 Mazbin, GNU Free Documentation License, Creative Commons Attribution - Share Alike 3.0 Unported.
Fig. 124 Roman gods next to a tricephalic entity with large lips. Dennevy, France.

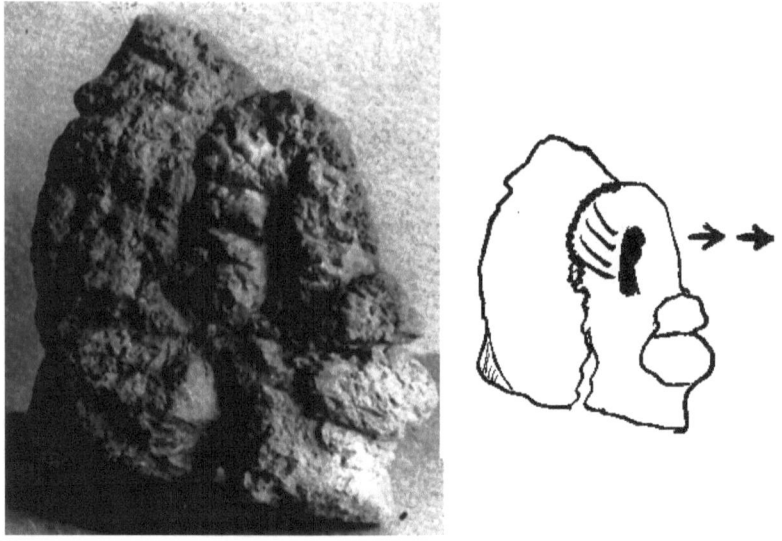

Fig. 125

Fig. 125 Sculpture of a human head with large lips. Height: 12.5 inches. Representation of a "man-fish" god. Campoligure (Genoa), Italy. P. Gaietto Collection.

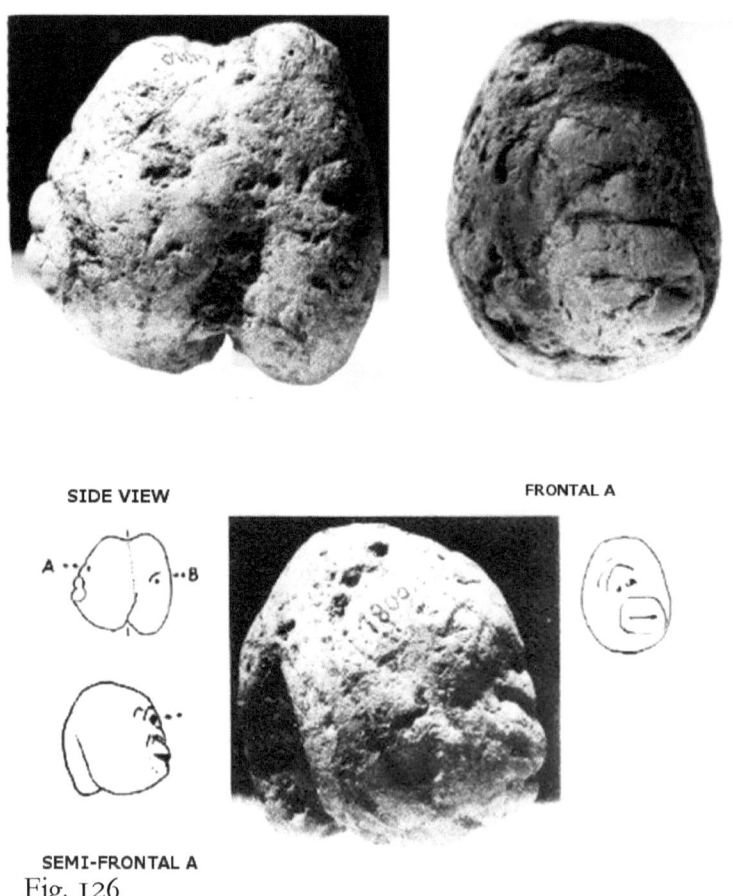

Fig. 126

Fig. 126 Sculpture of a human head with large lips, joined to a human head. Height: 2.5 inches. Representation of a "man-fish" god joined to another god. Genoa, Locality Sestri Ponente, slopes of Monte Gazzo. P. Gaietto Collection.

18

Snake

Snakes are all reptiles belonging to the suborder of scaly reptiles. They have no limbs and locomotion occurs with rapid undulations of the body. They live in warm areas and their head (Fig. 127) generally has the same shape in each species, regardless of their size.

In post-Paleolithic times, certain species of snakes were considered sacred animals in some civilizations, large or small, all over the world. In Western Asia, North Africa and Europe the snake was part of a cult, as shown in important works of religious art.

Among the ancient Egyptians, Meretseger was a goddess with a snake's head (Fig. 128), evidently a cobra, and protectress of the Thebes necropolis. This deity was also represented as a snake with a woman's head or as a sphinx with a snake's head, or even as a snake with three heads – of a woman, a snake and a vulture.

In Mesopotamia from 2,700-2,500 BC, a Babylonian god called Mingzida is shown in sculptures always holding up two large snakes.

In the city of Minos on Crete (1,500 BC) the "goddess of snakes" lifts up a snake in each hand, showing her breasts and wearing a beautiful dress.

Serapis was an Egyptian-Hellenistic deity whose cult was introduced into Egypt by Ptolemy I. His attributes were the Tricephalus (lion, wolf and dog) and the Serpent.

In central and southern Italy, the goddess Angizia was associated with a snake cult. A proto-historic divinity, its existence is documented in Abruzzo between the Neolithic and the Roman Age. The cult of this goddess survived into the Christian Age and was "Christianized" in Cocullo (province of L'Aquila) in the early 1400s. It is likely that the religious connotations of this goddess were replaced with those of Saint Dominic Abate. In Cocullo (897 m. L. M.) on the first of May there is the feast of San Domenico, whose carved effigy is wrapped in snakes (Fig. 129) and carried in a procession. The snakes remain on the statue if the temperature is cold; if the climate suddenly becomes hot, they crawl away because, being cold-blooded animals, they gain agility in heat.

The Neolithic is better known for its material culture (industries, animal domestication, vegetable cultivation) than its spiritual ones (art, religion); it began in South-West Asia several millennia earlier than in Europe.

A goddess-serpent was worshiped in shelter houses in 6,000 BC in the north of Greece.

There are to my knowledge no sculptures of snakes from the Neolithic, only engravings and reliefs of snakes from the Neolithic, only engravings and reliefs which generally represent coils of snakes which resemble decorations.

My hypothesis is that during the Paleolithic this animal, worshipped as a "sacred animal" and interpretable through sculpture, may have been there during the last interglacial, before the last glaciation. But since I never thought it could have been depicted, I never looked for it among the infinity of stone artefacts I selected over half a century.

It should be remembered that the Upper Paleolithic in Western Europe had a longer duration than in south-western Asia, and the spiritual culture of populations of these areas is known to us only from the art of the most important sites, where absolute dating was possible, and which generally represent single small civilizations. In the peripheral areas of these civilizations, finds of sculpture, probably from the same period, present difficulties for chronological interpretation.

The sculpture of a snake's head in Fig. 130 was found in the mountains of Liguria at Tiglieto (Genoa). It is 2.3 inches long and was found on the surface in a forest. It cannot be dated. It could belong to the Upper and Final Paleolithic as well as the Mesolithic or Neolithic, however only one find of a type is insufficient for an evaluation, but I thought it necessary to report it as it could be useful in the event of new discoveries.

In the evolutionist's interpretation, the mountain populations of Liguria had a tradition of sculpture that was absent in other areas.

Fig. 127

Fig. 128

Fig. 127 Head of a snake (python). Drawing from the end of the 19th century.
Fig. 128 Meretseger, a cobra-headed goddess of ancient Egypt. Copied from a fresco inside a pyramid.
Copyright 2013 A.Parrot, GNU Free Documentation License, Creative Commons Attribution - Share Alike 3.0 Unported.

Fig. 129

Fig. 130

Fig. 129 A sculpture of San Domenico Abate wrapped in snakes and carried in a procession. Christian religious festival, Catholicism.
Fig. 130 A sculpture of a snake's head. Length: 2.3 inches. Tiglieto (Genoa), Italy. P. Gaietto Collection.

Publications and cultural presentations by the same author

"L'arte nasce agli albori del Quaternario" (Sabatelli, Savona, 1968)
"Arte vergine" (C.S.I.O.A., Genova, 1974)
"Favola itinerante dell'uomo dell'Età della Pietra in Liguria" (G. & G. Del Cielo, Genova, 1976)
"Prescultura e scultura preistorica" (E.R.G.A., Genova, 1982)
Une sculpture zoomorphe suspendue du Mousterien (Congrès International de Paléontologie humaine, Nice, 1982)
Une sculpture anthropomorphe aux deux faces du Surrey (Primeval Sculpture I, Primigenia, 1984)
Un gisément moustérien sans art au Liban (ibidem)
The anthropomorphic double-faced divinity in the sculpture of the Lower and Middle Paleolithic (Primeval Sculpture II, Primigenia, 1984)
To be or not to be: that is the question (Primeval Sculpture III, Primigenia, 1984)
Fondazione e direzione del Museo delle Origini dell'Uomo (www.museoorigini.it, 2000)
Il volto megalitico di Borzone (Paleolithic Art Magazine, www.paleolithicartmagazine.org, 2000)
L'abbigliamento nelle "Veneri" di Liguria, Austria e Messico (ibidem)
L' intuizione di Boucher de Perthes (ibidem)
L'urlo di Homo Erectus (ibidem)
L'antica ceramica zooantropomorfa del Messico in relazione alle "Veneri" bifronti paleolitiche dei Balzi Rossi (ibidem)
Il bifrontismo con gli uccelli (Paleolithic Art Magazine, 2001)
Una scultura litica zooantropomorfa bifronte dell'Acheuleano evoluto di Roma-Torre in Pietra interpretata attraverso la tipologia delle sculture (ibidem)
L'origine dell'arte decorativa è nell'Acheuleano (ibidem)
Gli utensili litici e gli utensili lignei per la fabbricazione di utensili e sculture litiche nell'Acheuleano (ibidem)
Arte e Paletnologia (ibidem)
Una scultura litica zooantropomorfa bifronte dell'Acheuleano evoluto dell'Italia meridionale (ibidem)
Una scultura litica antropomorfa bifronte del Paleolitico inferiore della Danimarca (ibidem)
I cibi artistici rituali in Italia, da *Homo Erectus* a *Homo Sapiens Sapiens* (ibidem)
Aspetti della cultura materiale e della cultura spirituale nelle scoperte dell'Archeoastronomia e loro inserimento nella Paletnologia (Convegno Internazionale di Studi Liguri, 2002)
Breve storia delle scoperte dell'arte del Paleolitico inferiore, e ipotesi sul futuro della ricerca (Paleolithic Art Magazine, 2002)
L'idolatria nei colossi antropomorfi paleolitici e post-paleolitici (ibidem)
Affinità tra la Venere paleolitica con due teste dei Balzi Rossi (Liguria) e la Venere neolitica con due teste di Campo Ceresole (Lombardia)
Le Erme quadrifronti di Roma (ibidem)
Un ritratto umano scolpito 200.000 anni fa descritto con la didattica dell'Arteologia (ibidem)
Il colosso di Whangape della Nuova Zelanda attribuito al Paleolitico superiore (ibidem)
Una doppia statuina antropomorfa del Paleolitico superiore (Paleolithic Art Magazine, 2007)
Definizione degli studi sull'arte del Paleolitico inferiore e medio (Paleolithic Art Magazine, 2009)
Erotic Art? ("100.000 years of Beauty", Gallimard, Paris, 2009)
Phylogenesis of Beauty, 2008 (www.Lulu.com)
Intelligent cells and their inventions, 2010 (www.Lulu.com)
Erotism and religion, 2011 (www.Lulu.com)
Anthropomorphic Paleolithic Sculpture, 2012 (www.Lulu.com)
Catalogo della scultura paleolitica europea, 2012 (www.Lulu.com)
Gli animali sacri nella scultura del Paleolitico, 2013 (www.Lulu.com)
An Iconography of Western Religions, 2013 (www.Lulu.com)
Concettuario degli stili, 2016 (www.Lulu.com)
Horse and Wheel, 2016 (www.Lulu.com)
Dog and Man, 2017 (www.Lulu.com)
Origin of Man Science and prehistoric art, 2017 (www.Lulu.com)
Caccia e gastronomia, 2018 (www.Lulu.com)
La molletta pinzante, 2018 (www.Lulu.com)
Asce, 2018 (www.Lulu.com)

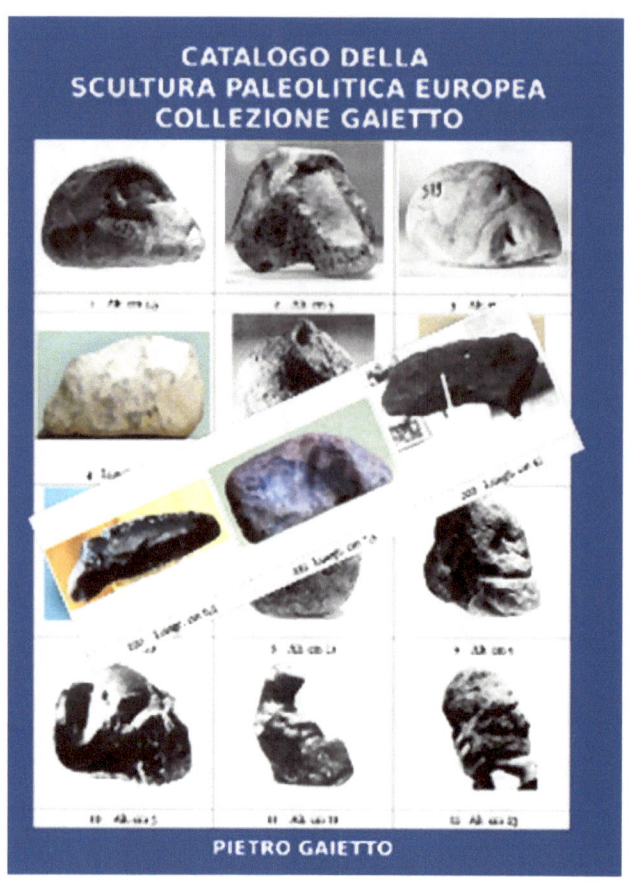

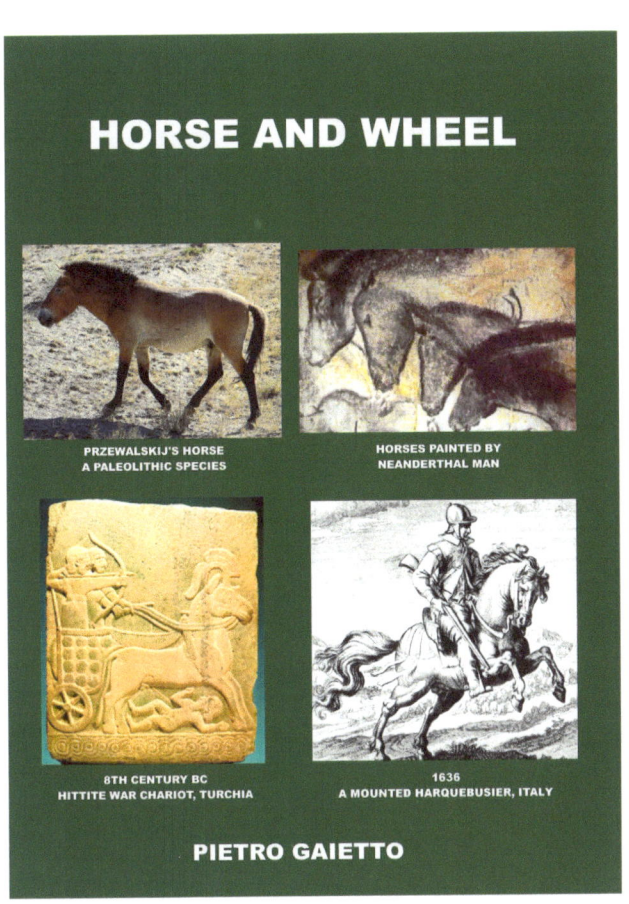

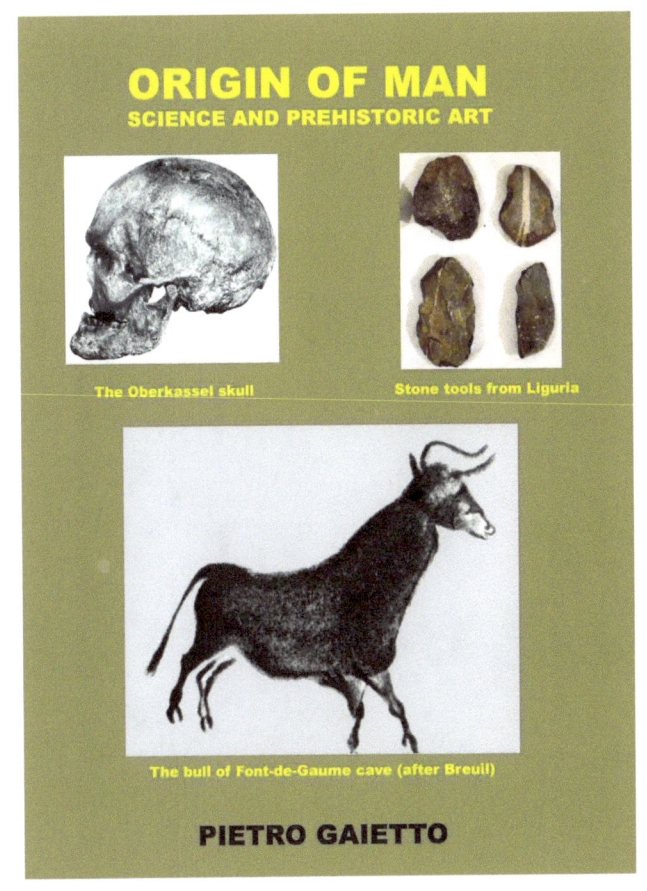

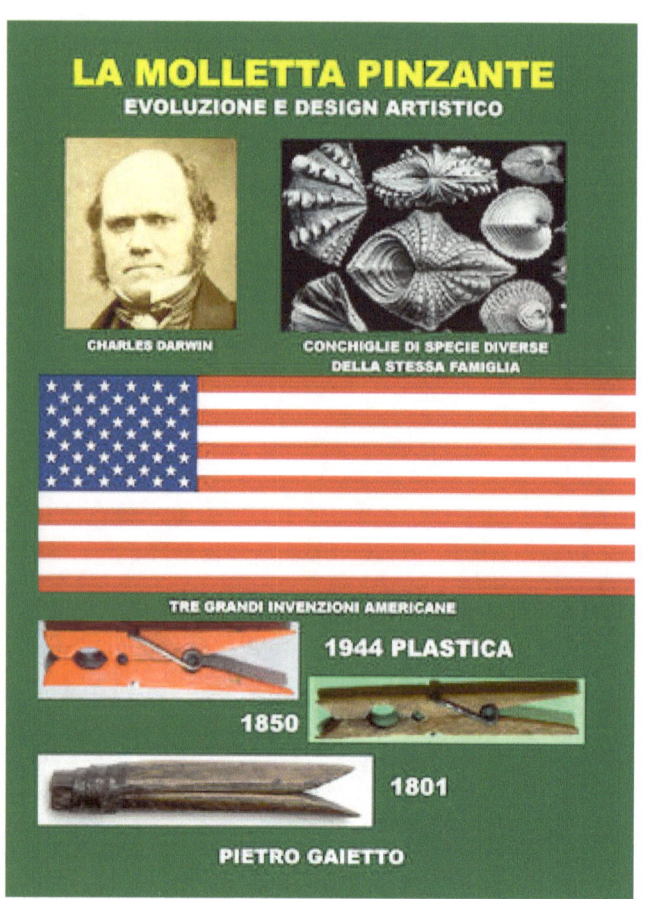

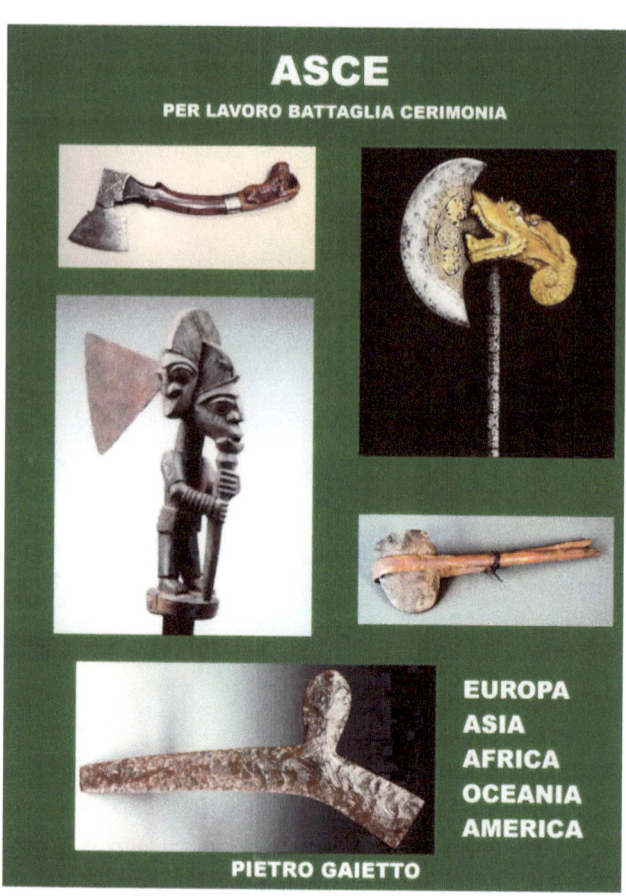

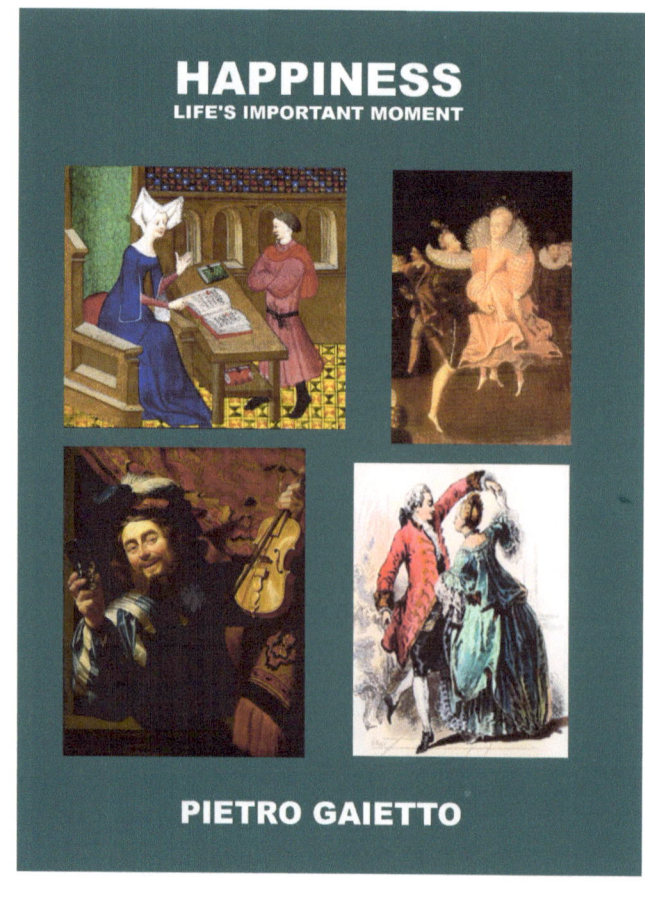

www.ingramcontent.com/pod-product-compliance
Lightning Source LLC
Chambersburg PA
CBHW050942200526
45172CB00020B/530